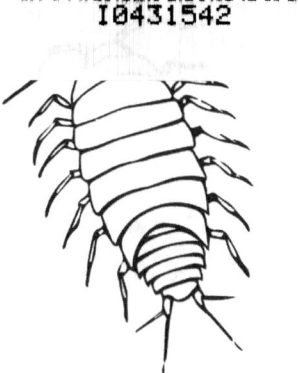

FOR THE LOVE OF BUGS

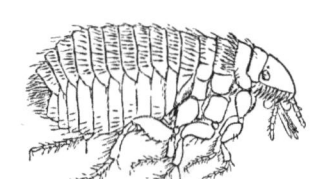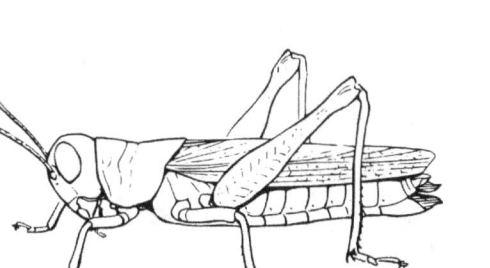

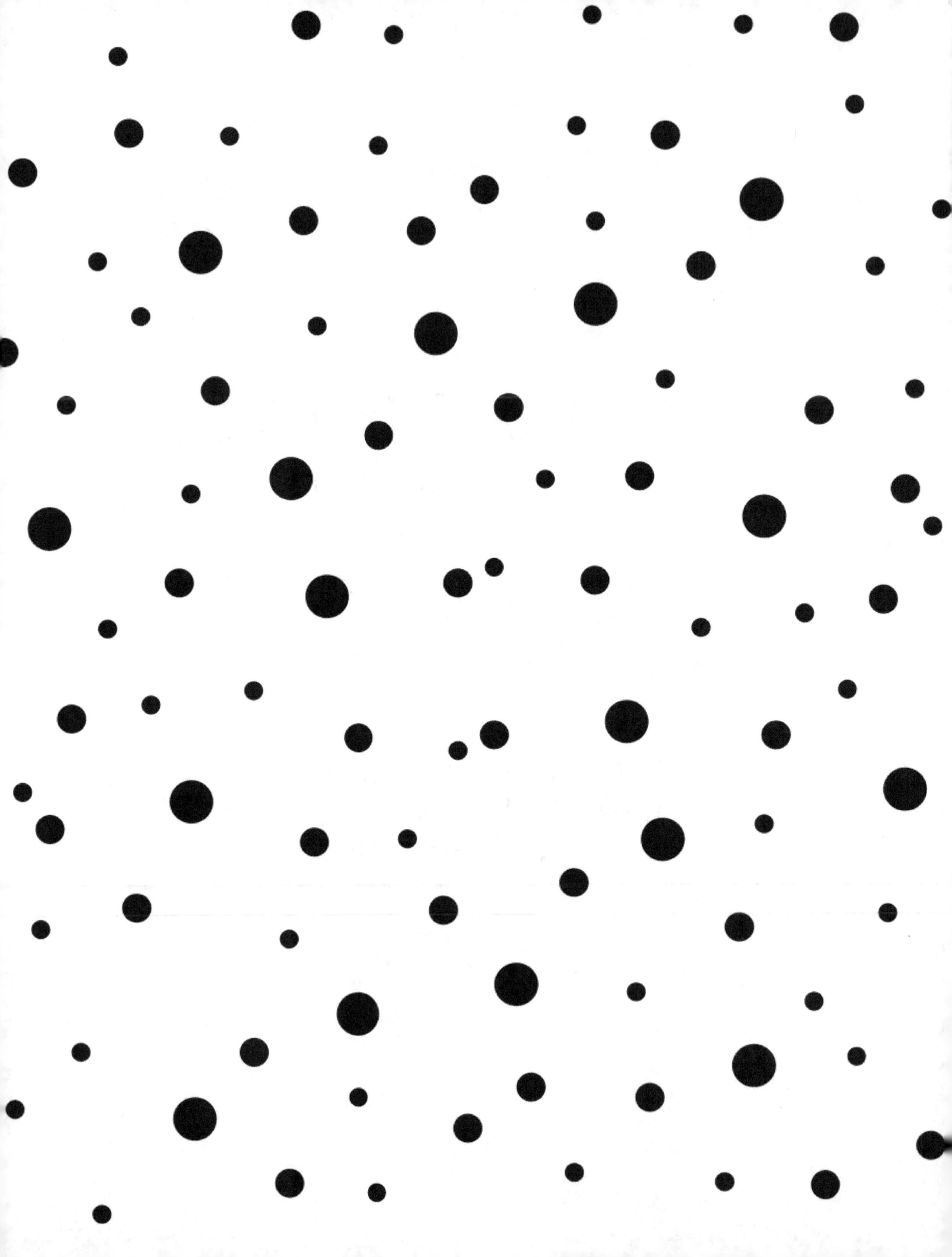

This book belongs to:

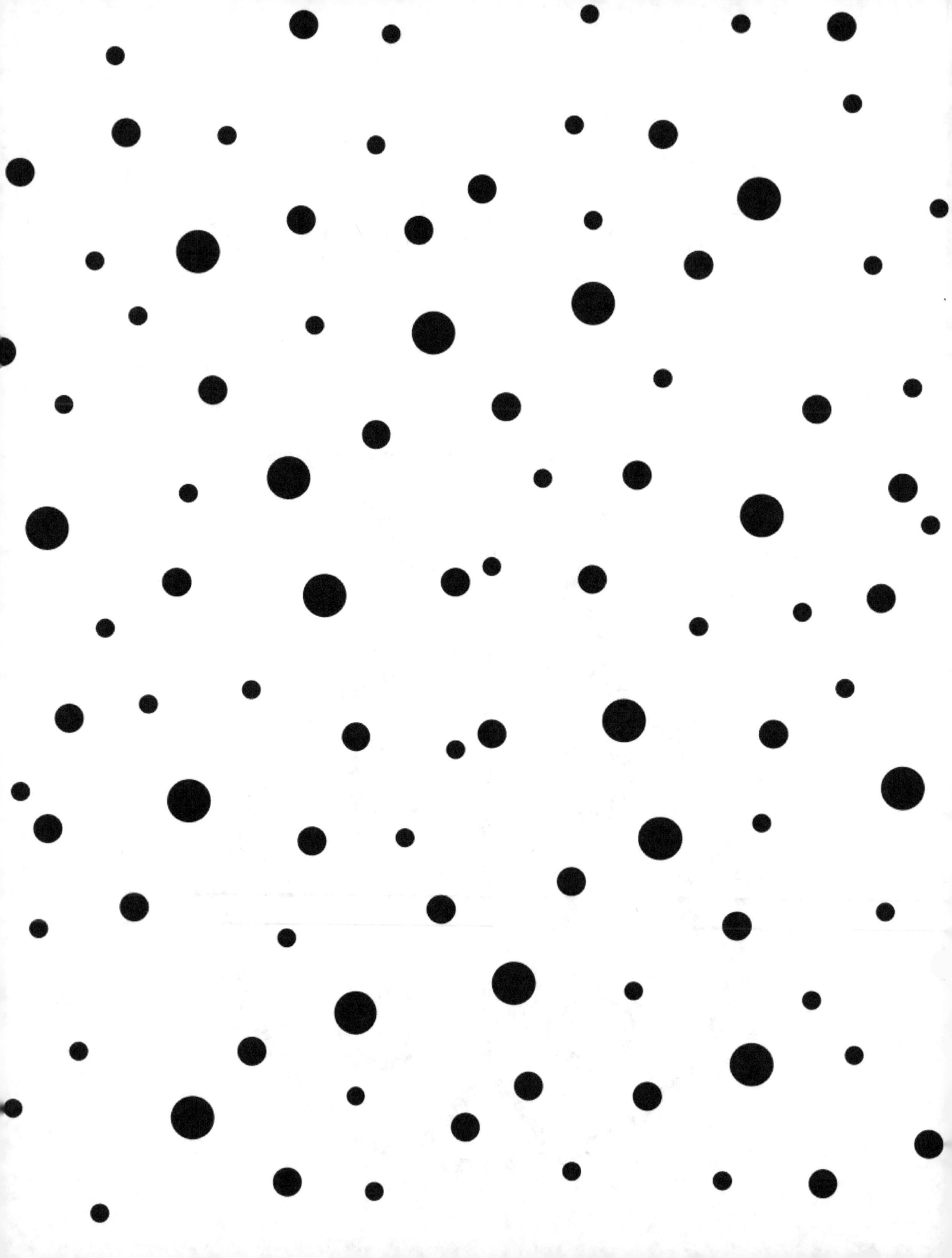

TRIVIA TIME -

TAKE A BREAK FROM REALITY!

Hello Gorgeous!

Thank you for purchasing one of our books. We appreciate your kindness.

We hope that this book will bring you a sense of peace of mind and relaxation as you go through the pages with your favorite crayons, pens, pencils are markers.

About the Author

Litasha Greaves is the creative mind and president of Trivia Time. She is a mother of two boys and two girls, a wife, a Jesus lover and follower and a fun-fanatic.

Trivia Time is an organization built on the president,s love of having fun and playing games. She is dubbed Quiz Master by many of her colleagues and family.
Today, Trivia Time has a YouTube Channel where quiz and trivia activities are shared with the world. The organization has also taken a step to prove more fun content to its audience through coloring books, puzzle books and workbooks for kids.

Please check us out on Youtube @ Trivia Time

I would love to hear from you. Leave a review on Amazon.

Copyright © 2022 by Litasha Greaves at Trivia Time. All rights reserved.

No part of this book may be reproduced or used in any manner without the prior written permission of the copyright owner, except for the use of brief quotations in a book review.

To request permission, contact publisher at litashagreaves@yahoo.com

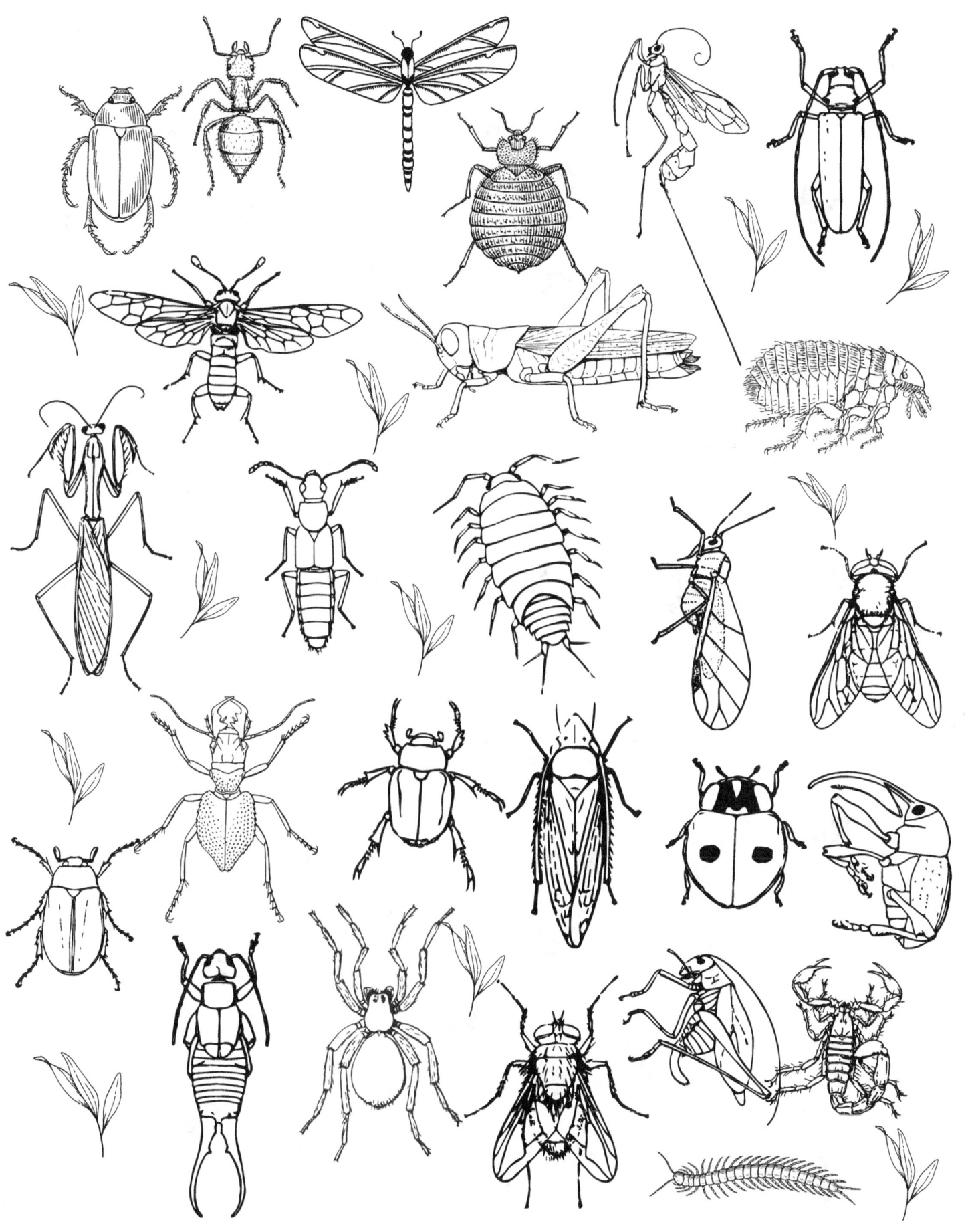

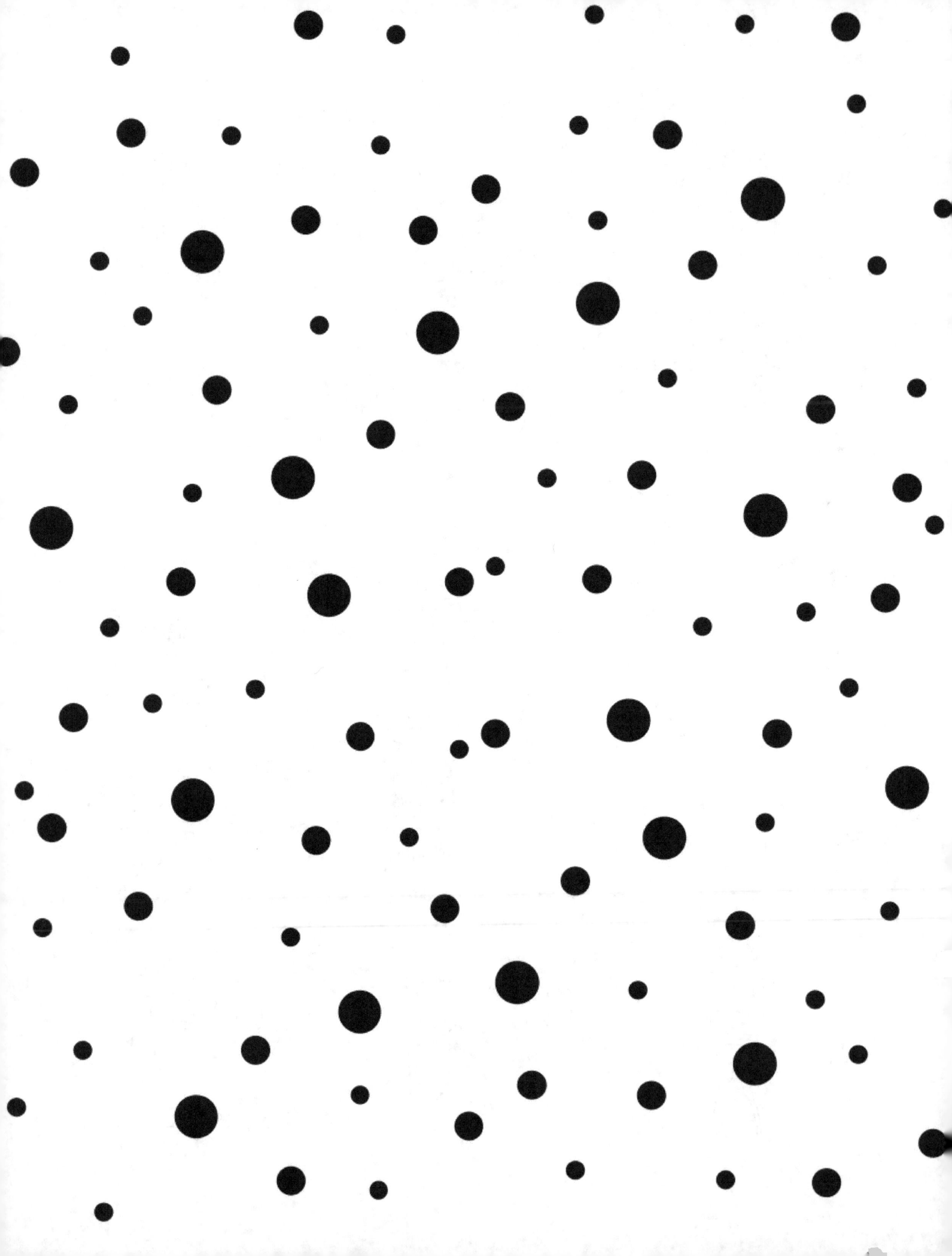

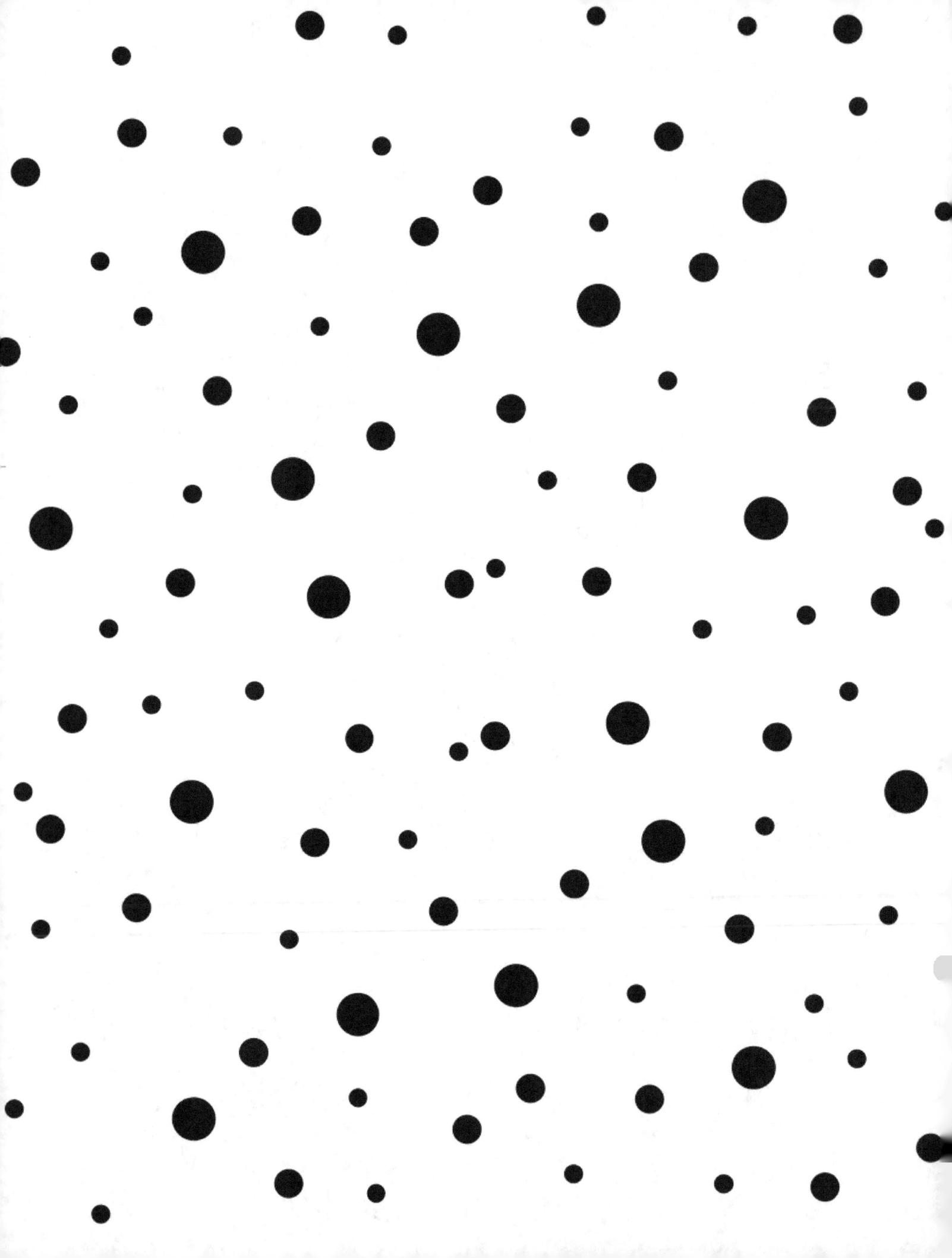

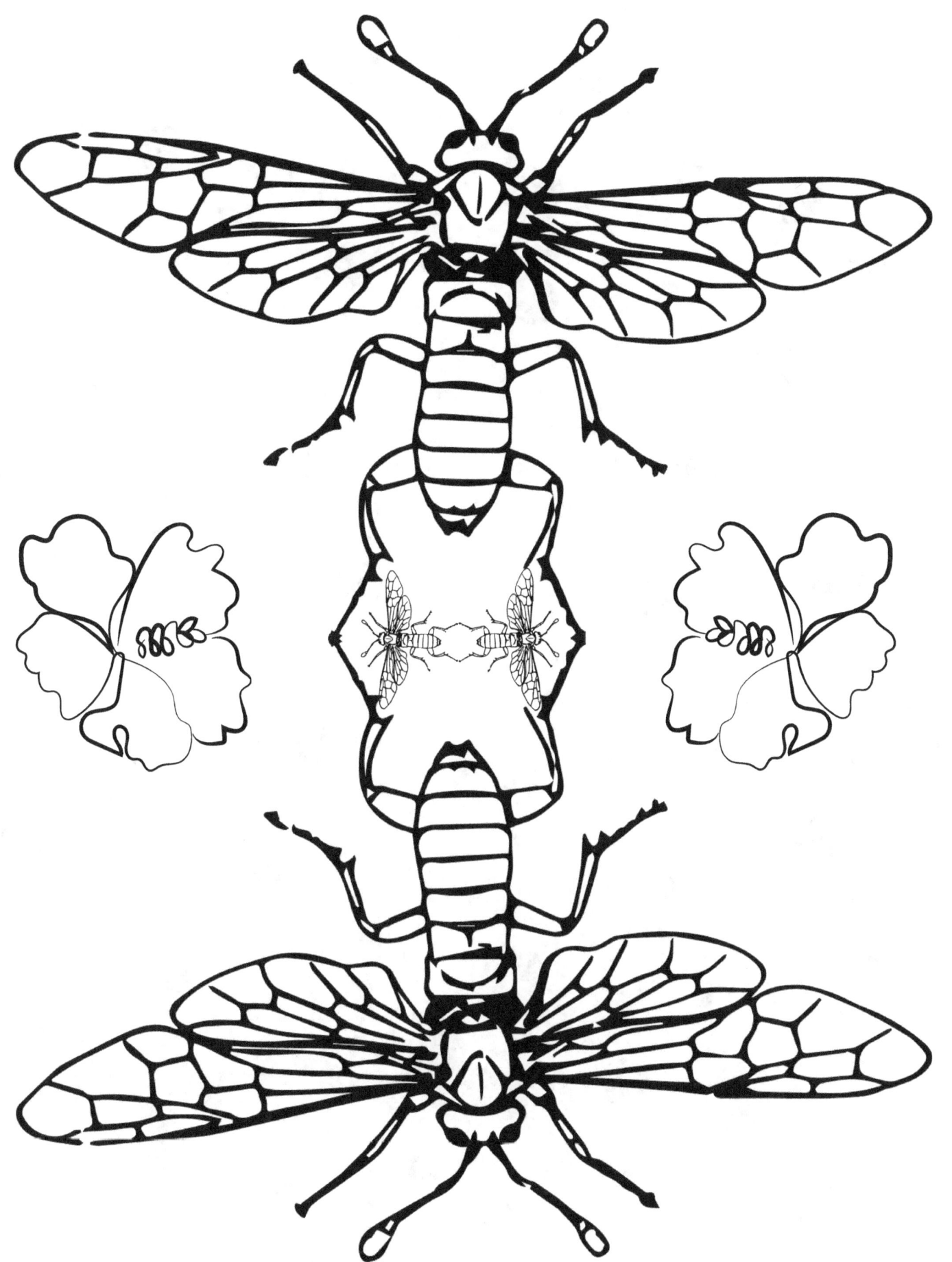

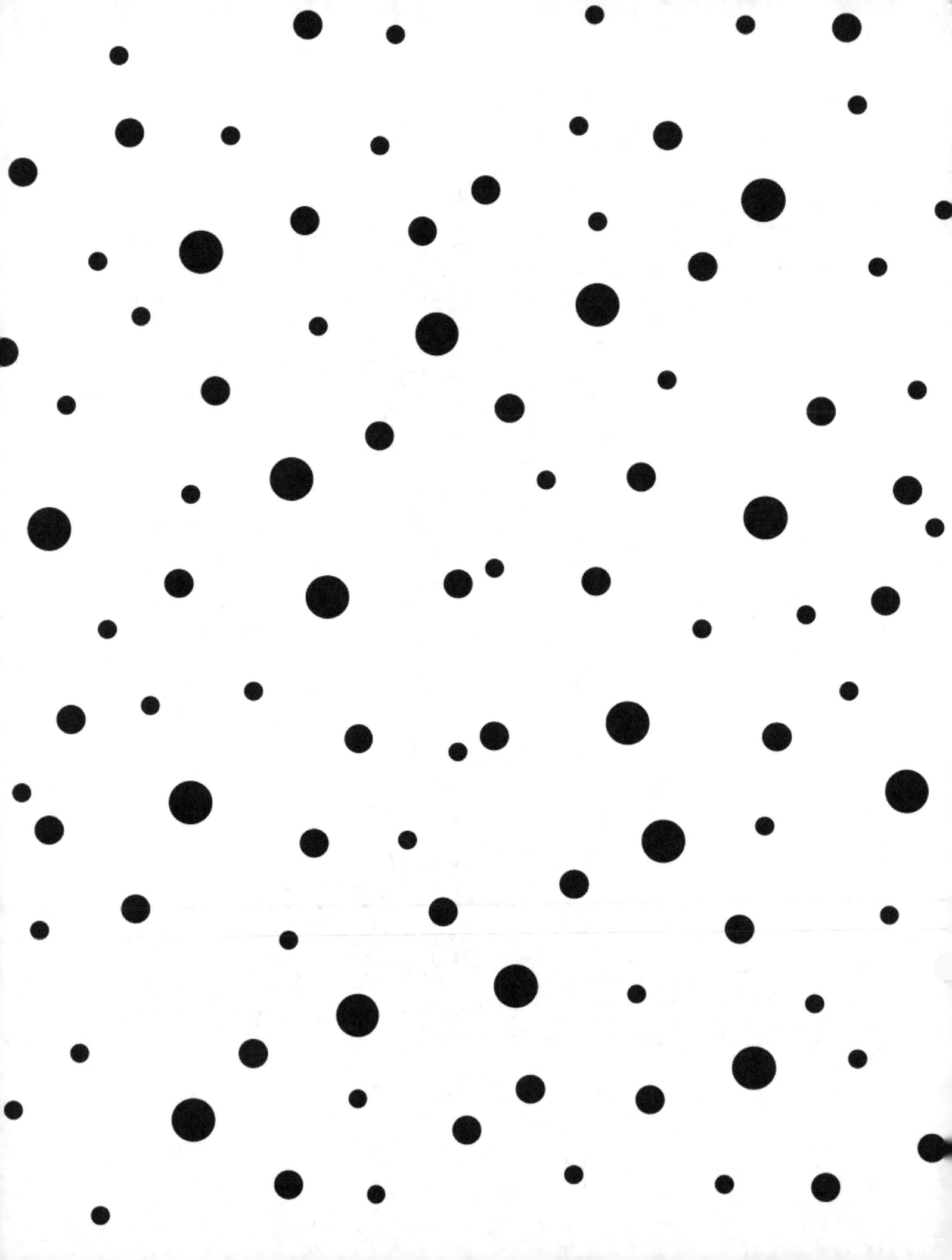

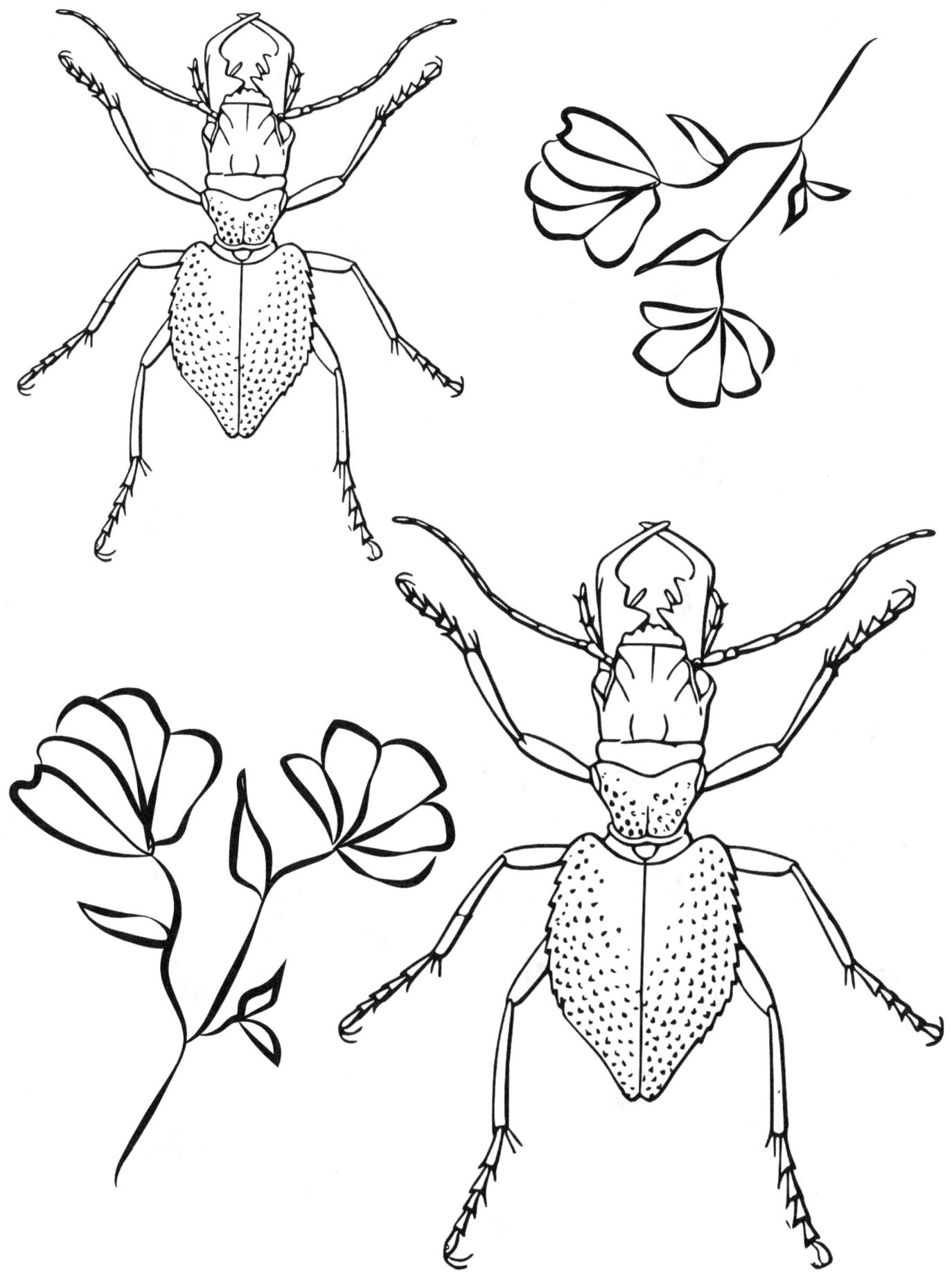

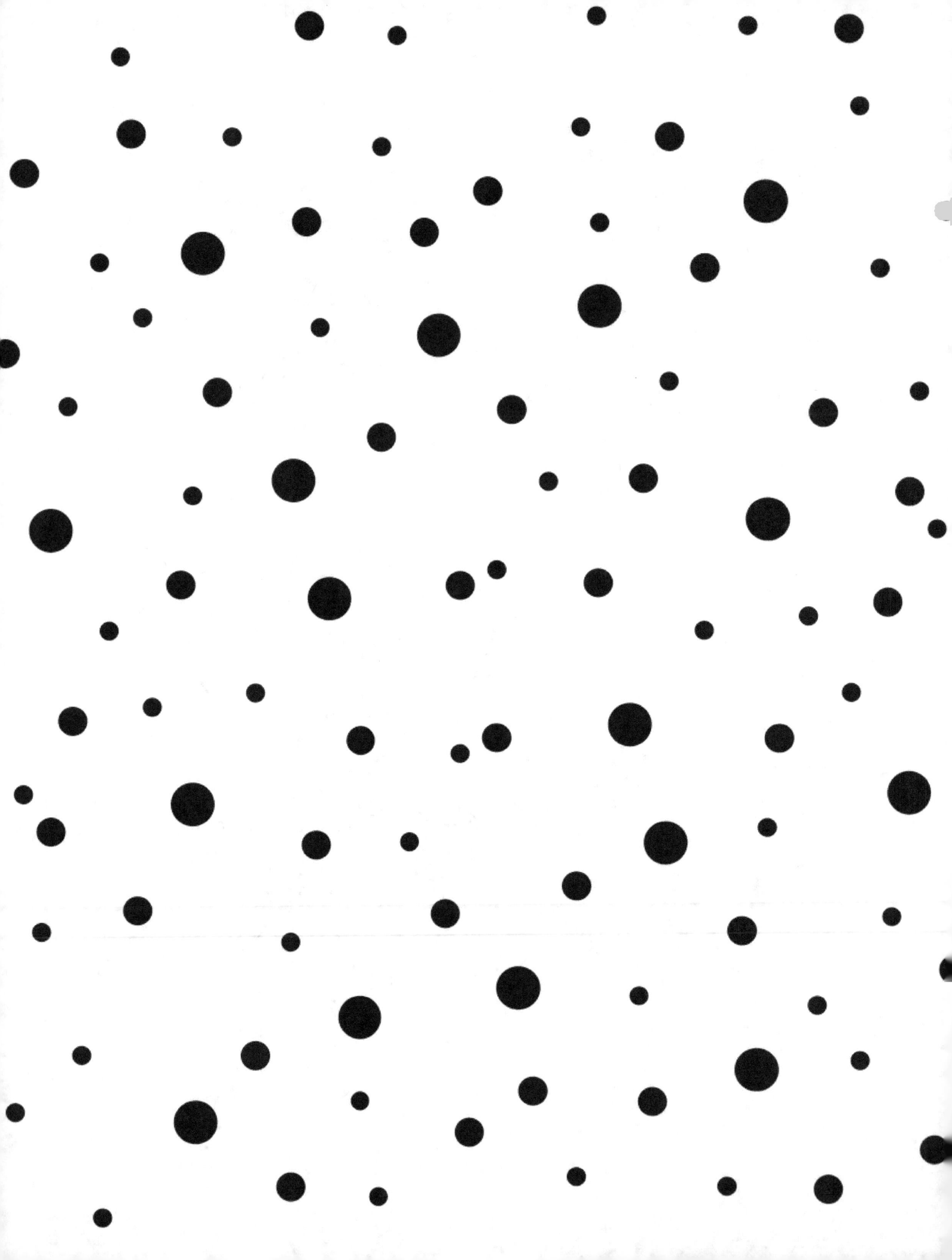

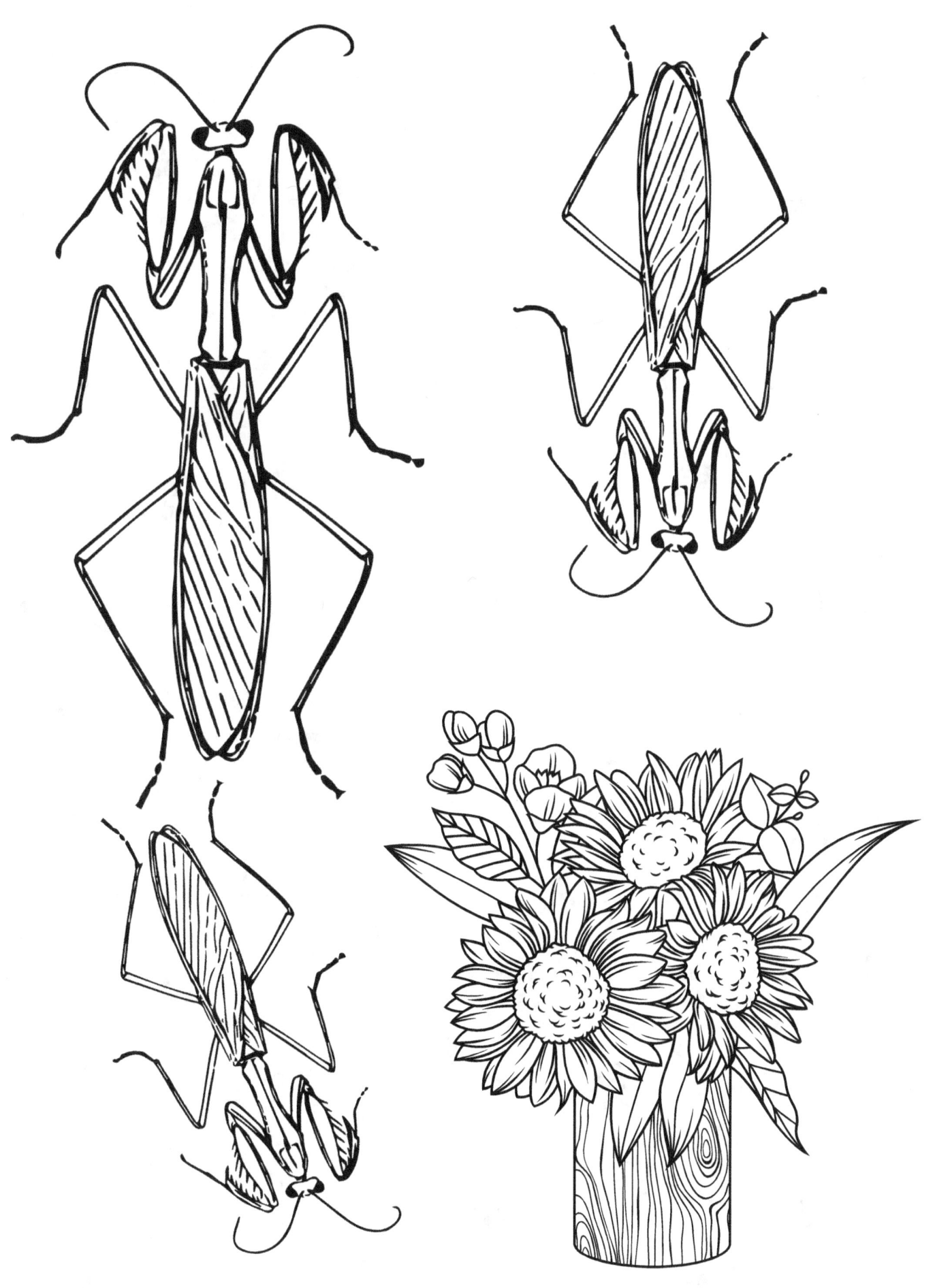

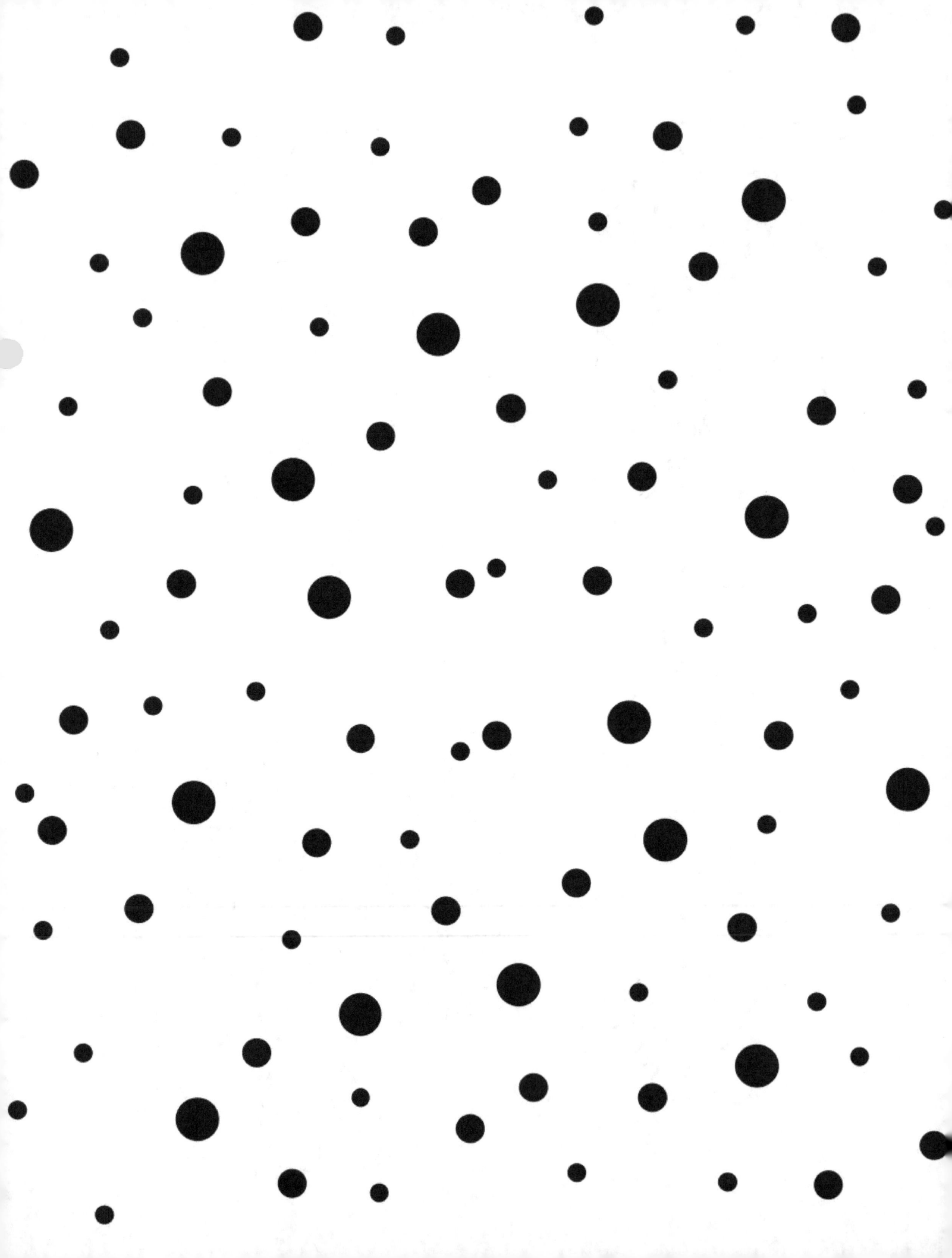

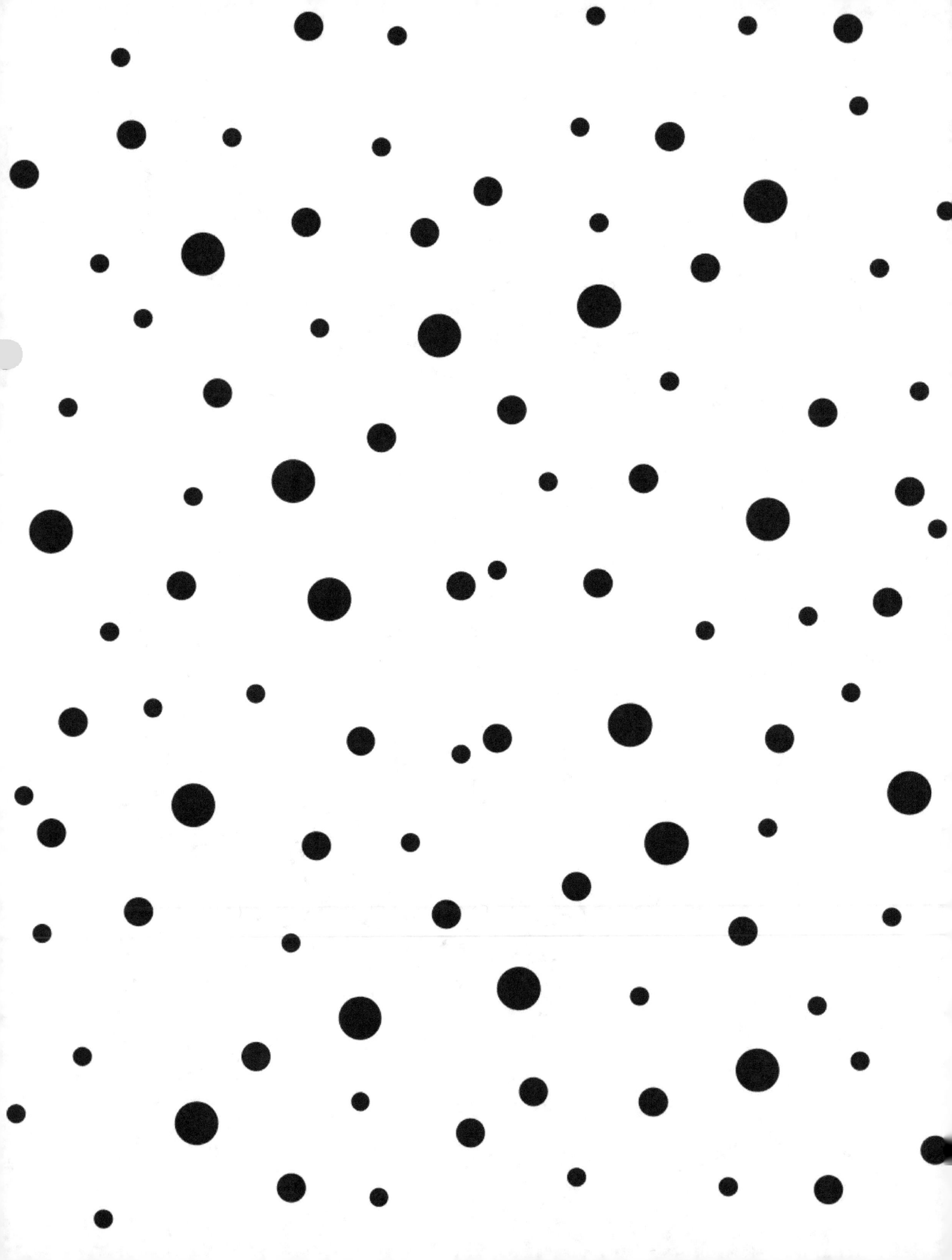

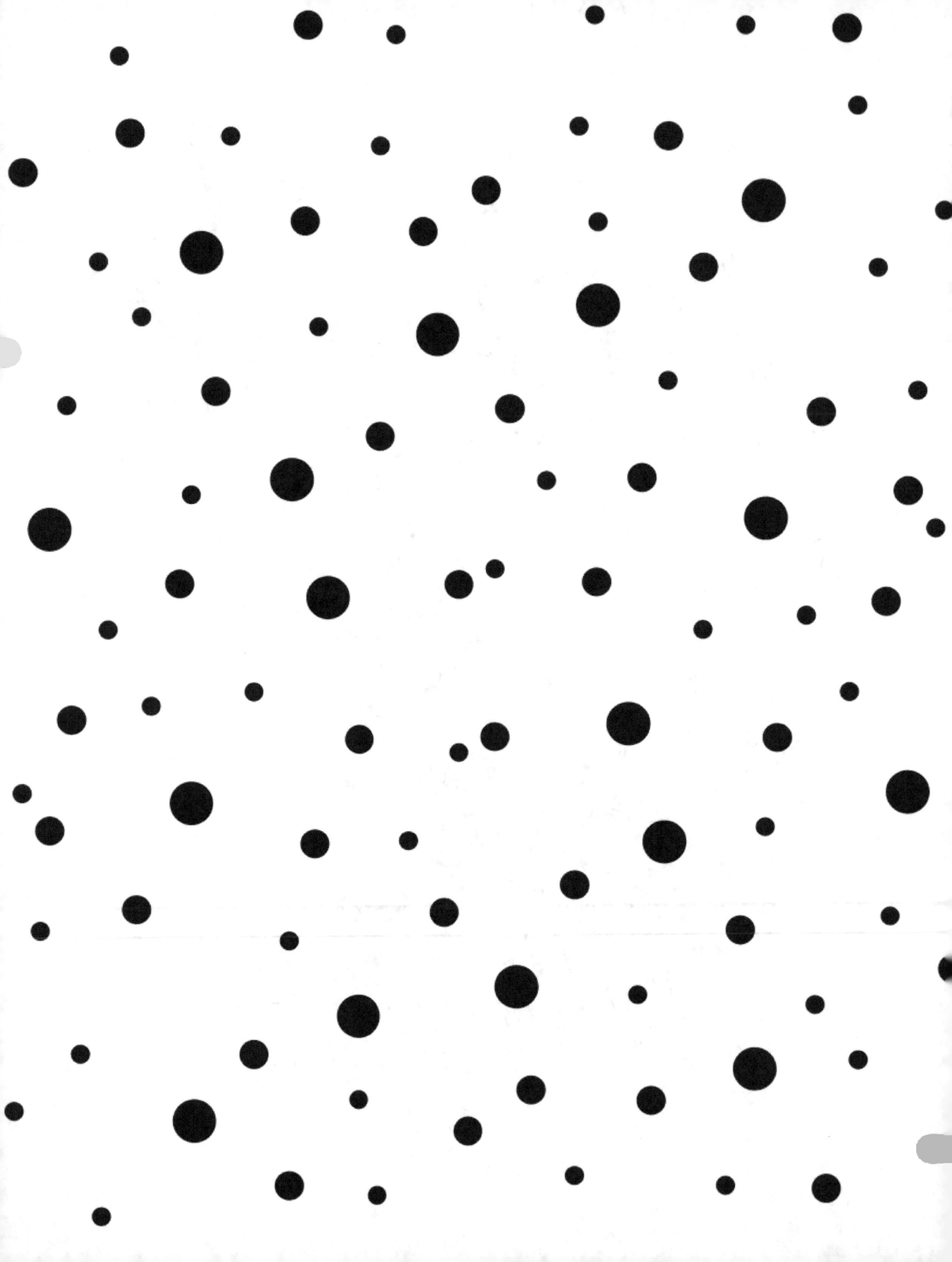

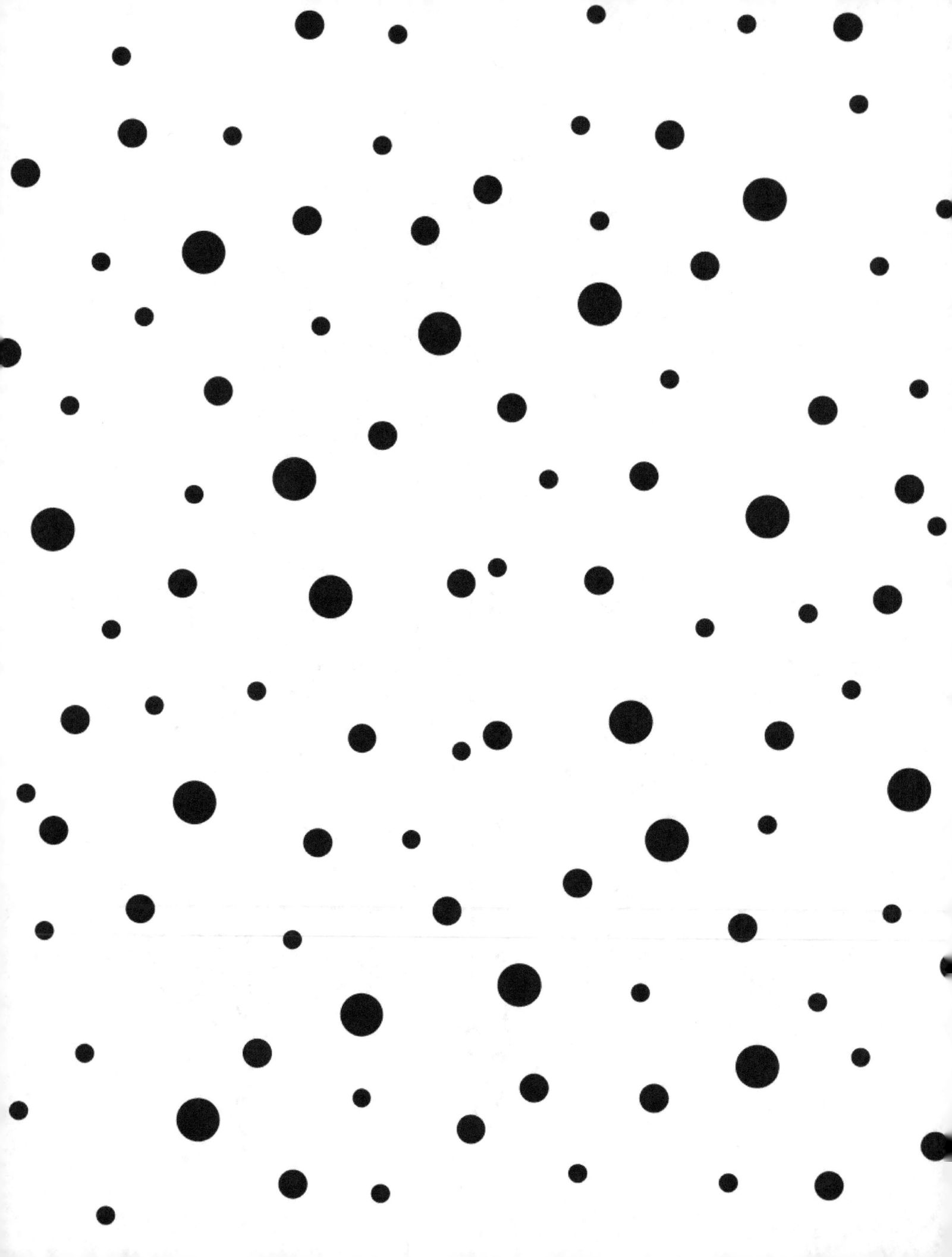

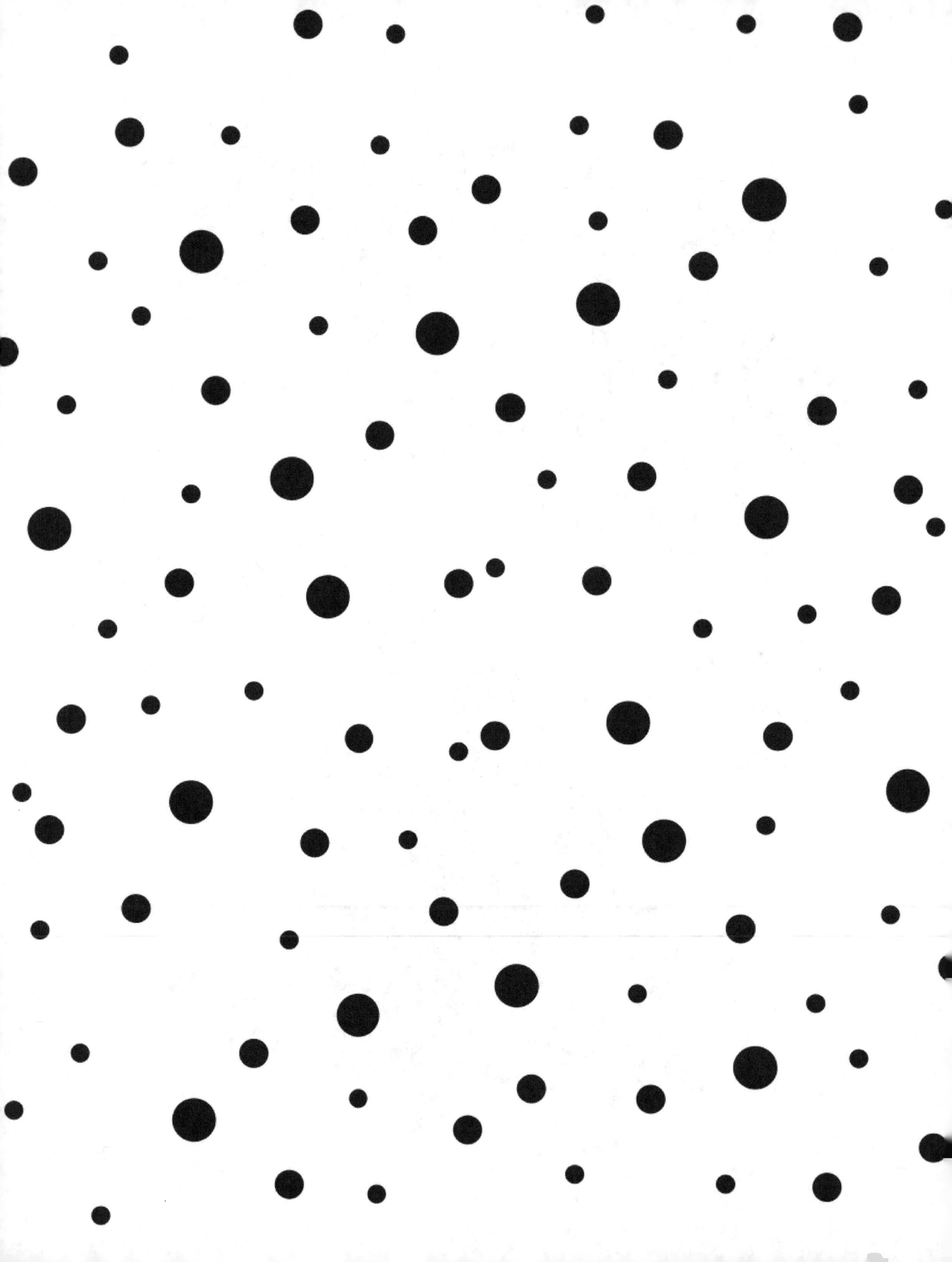

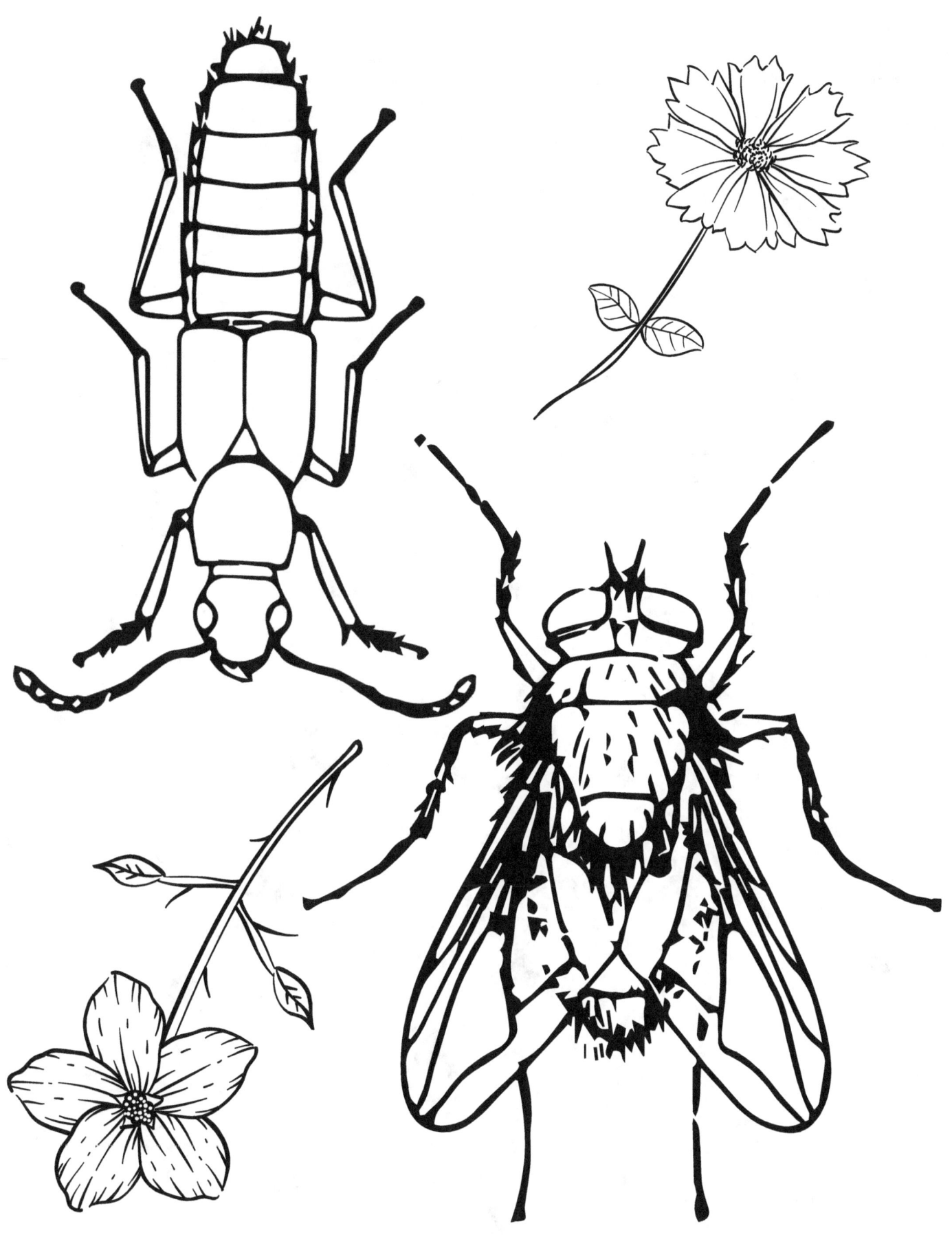

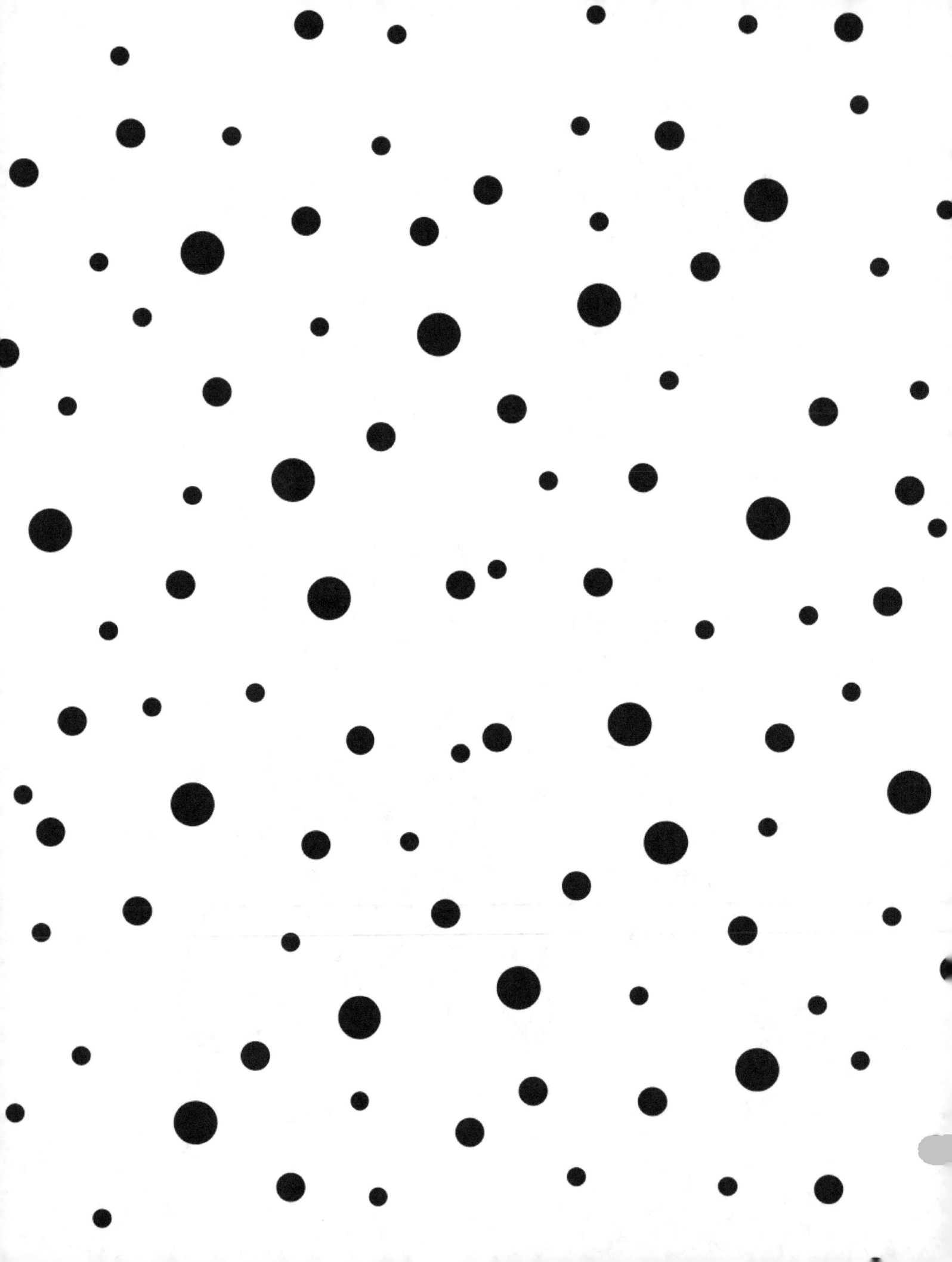

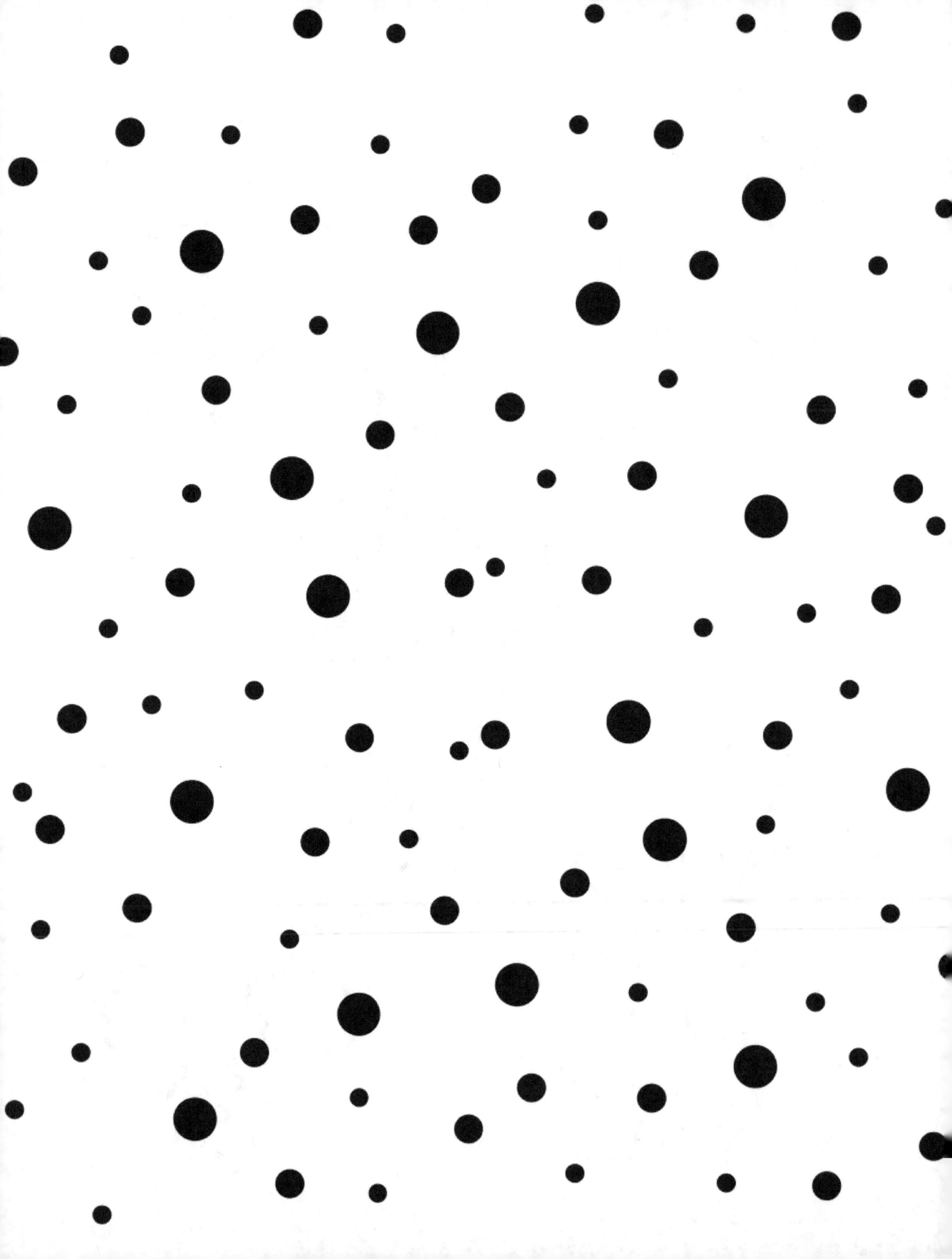

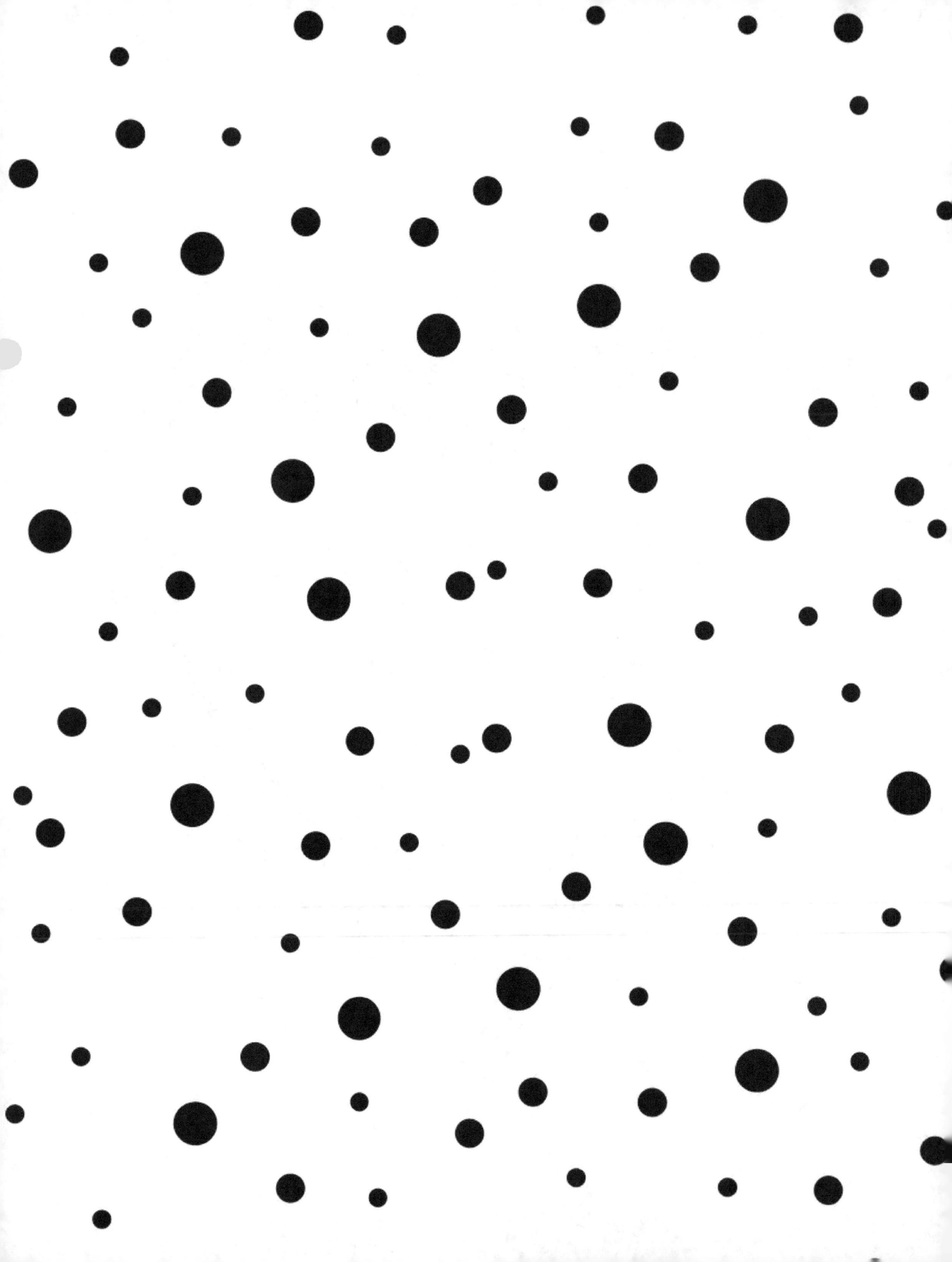

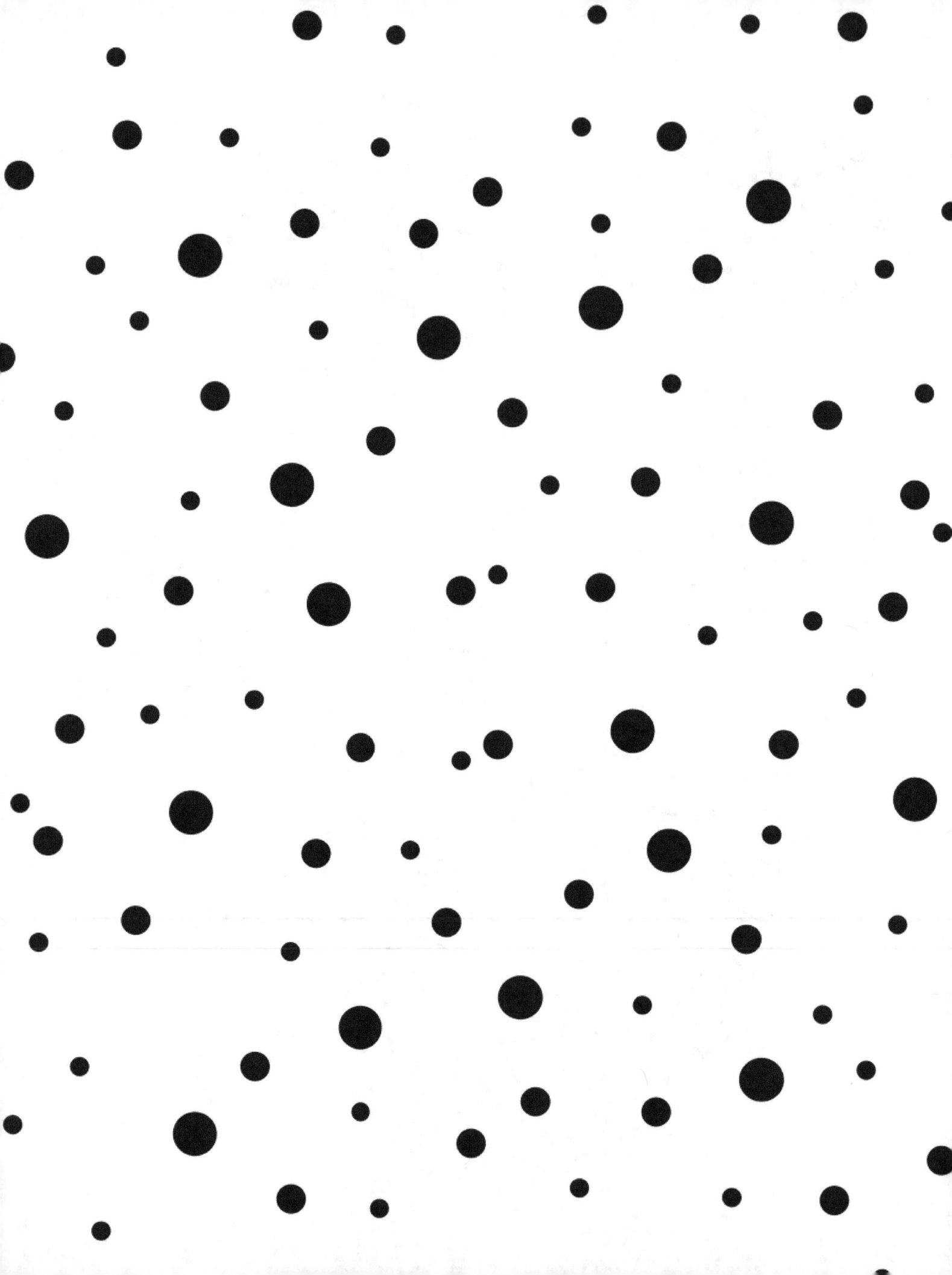

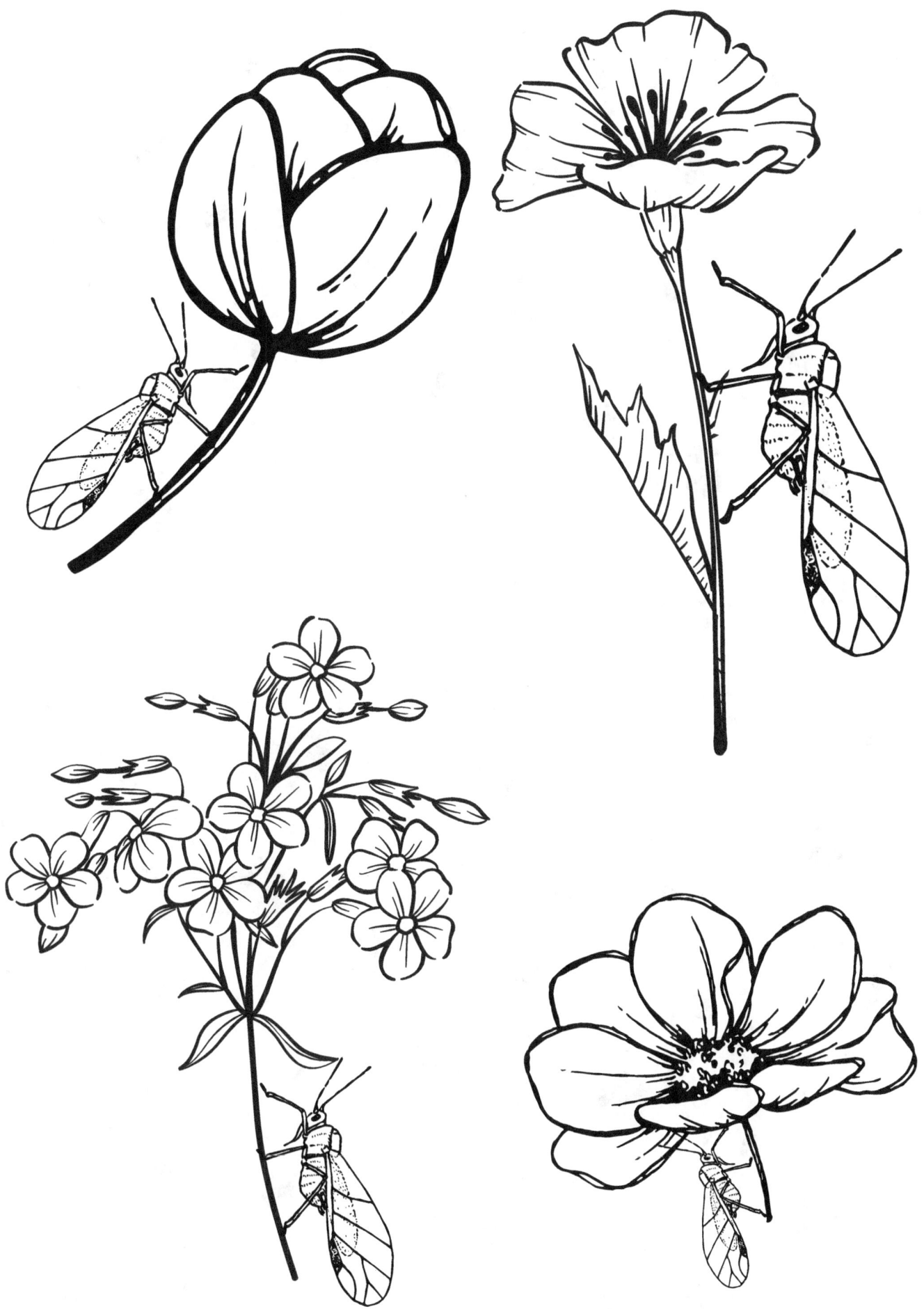

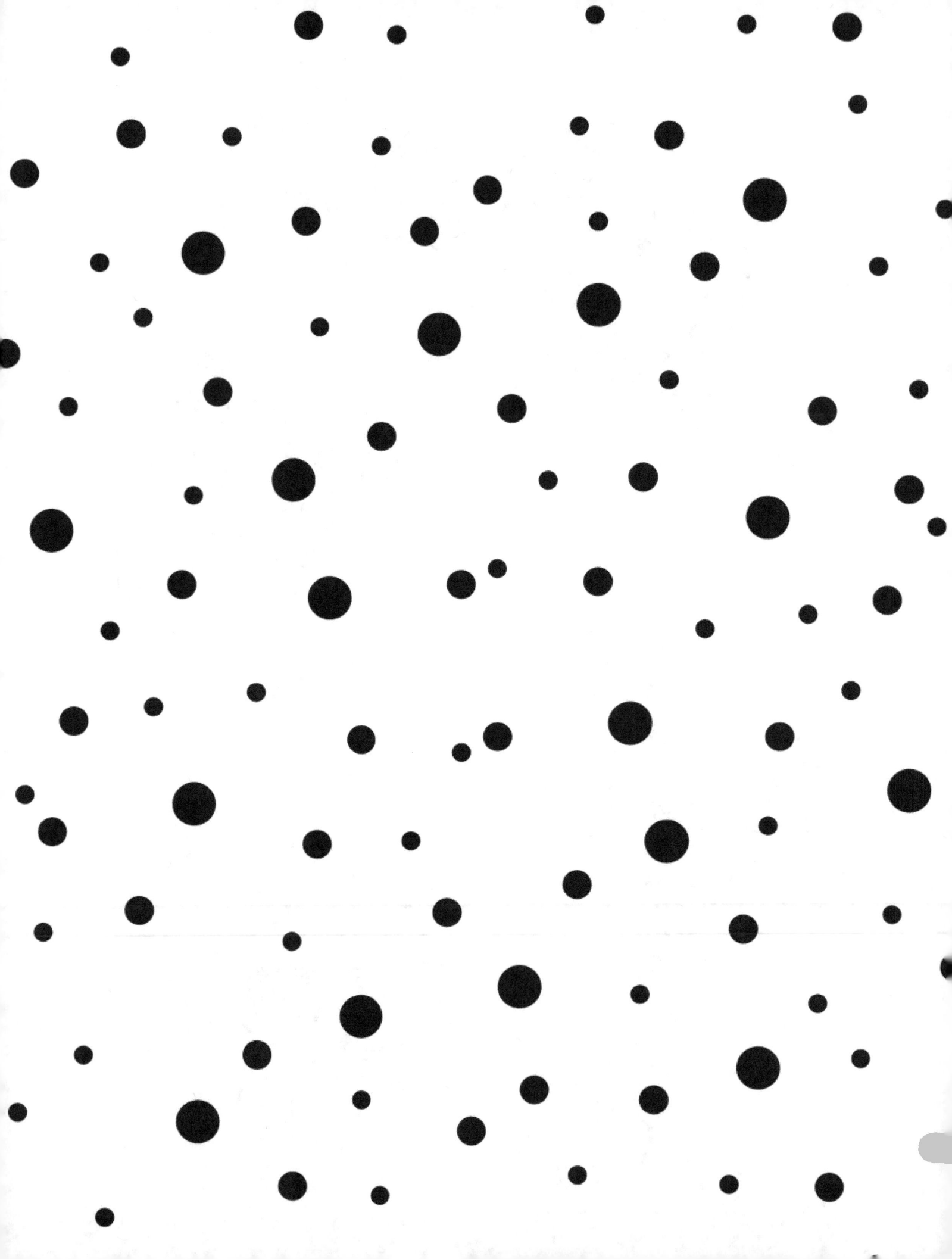

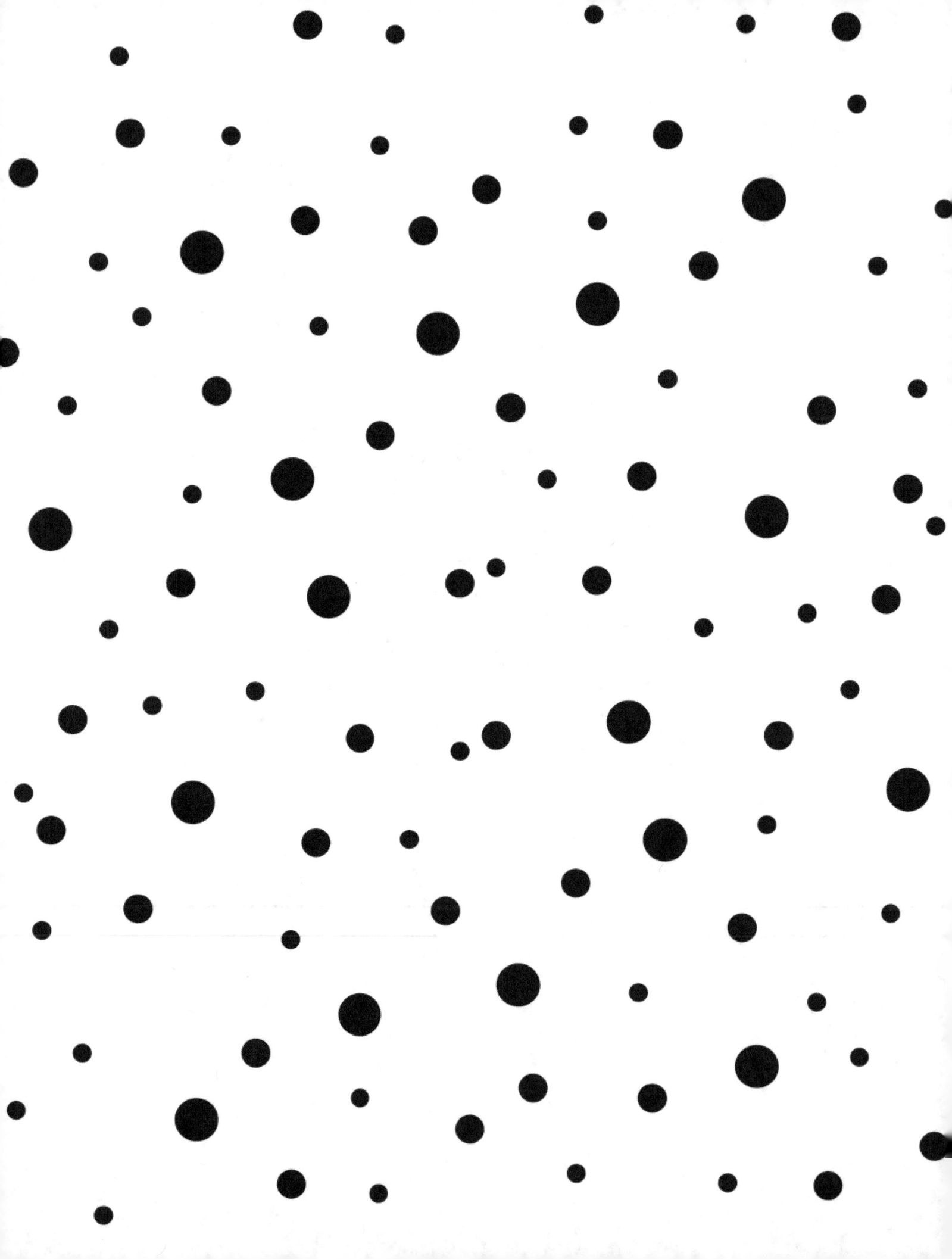

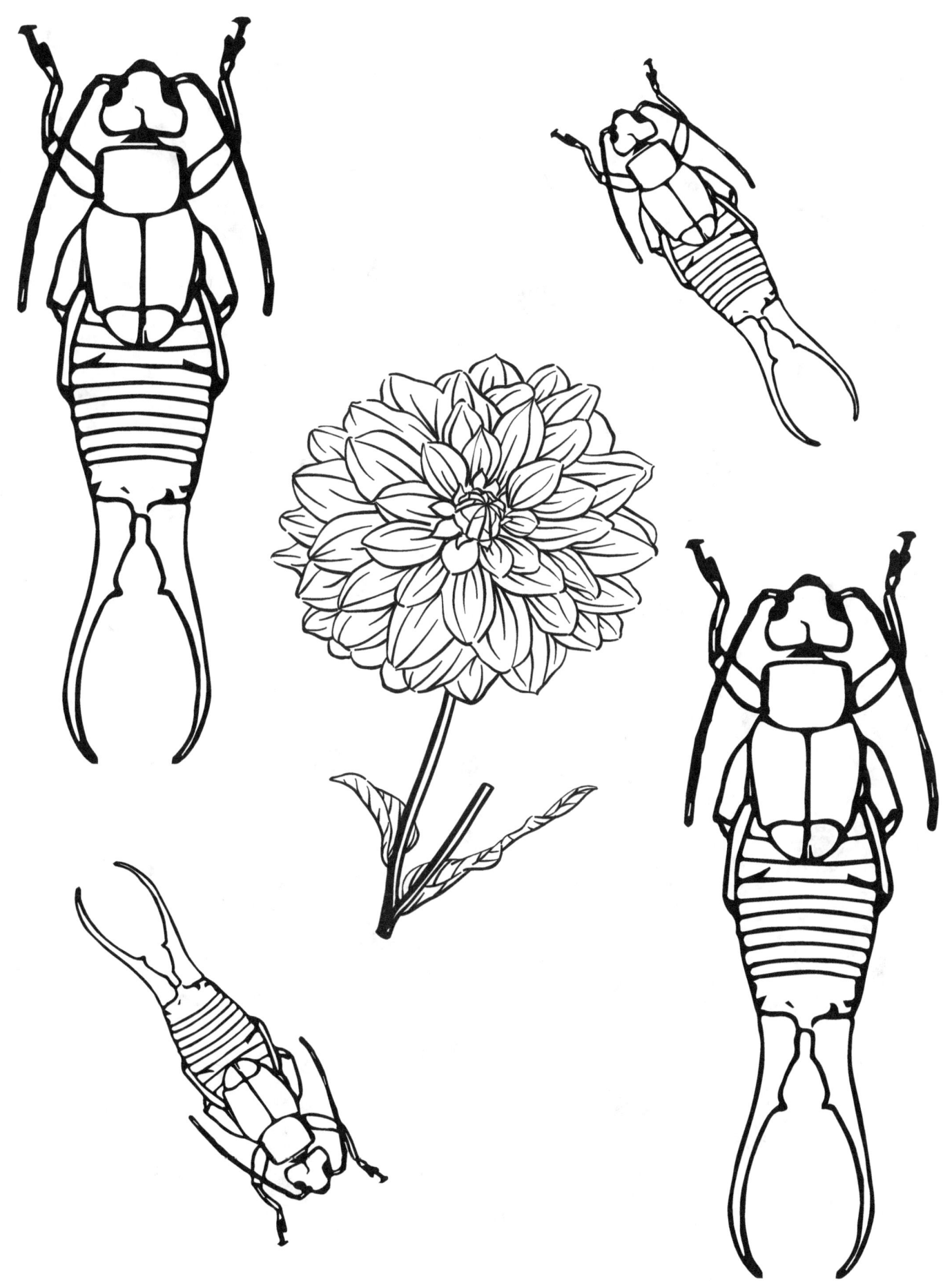

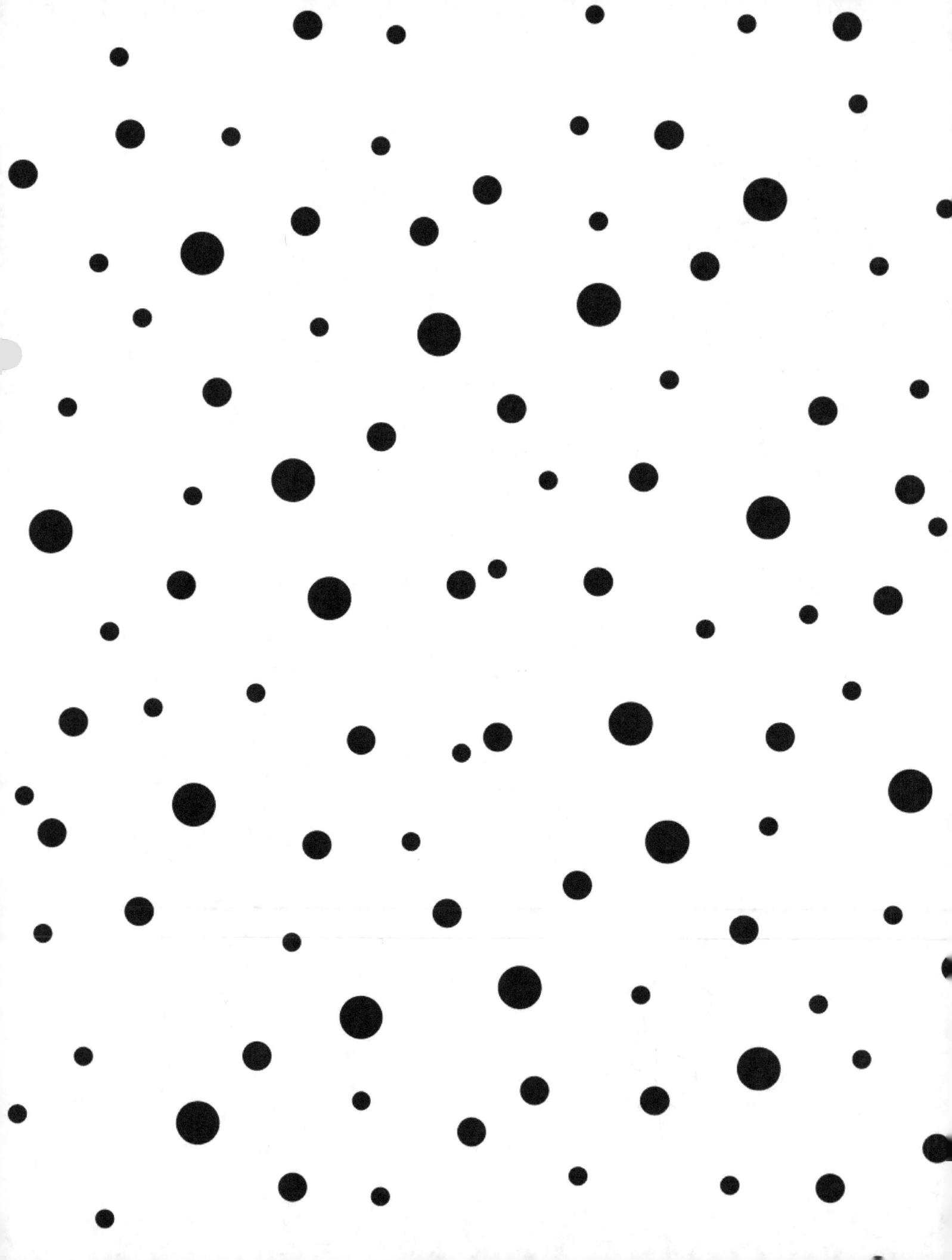

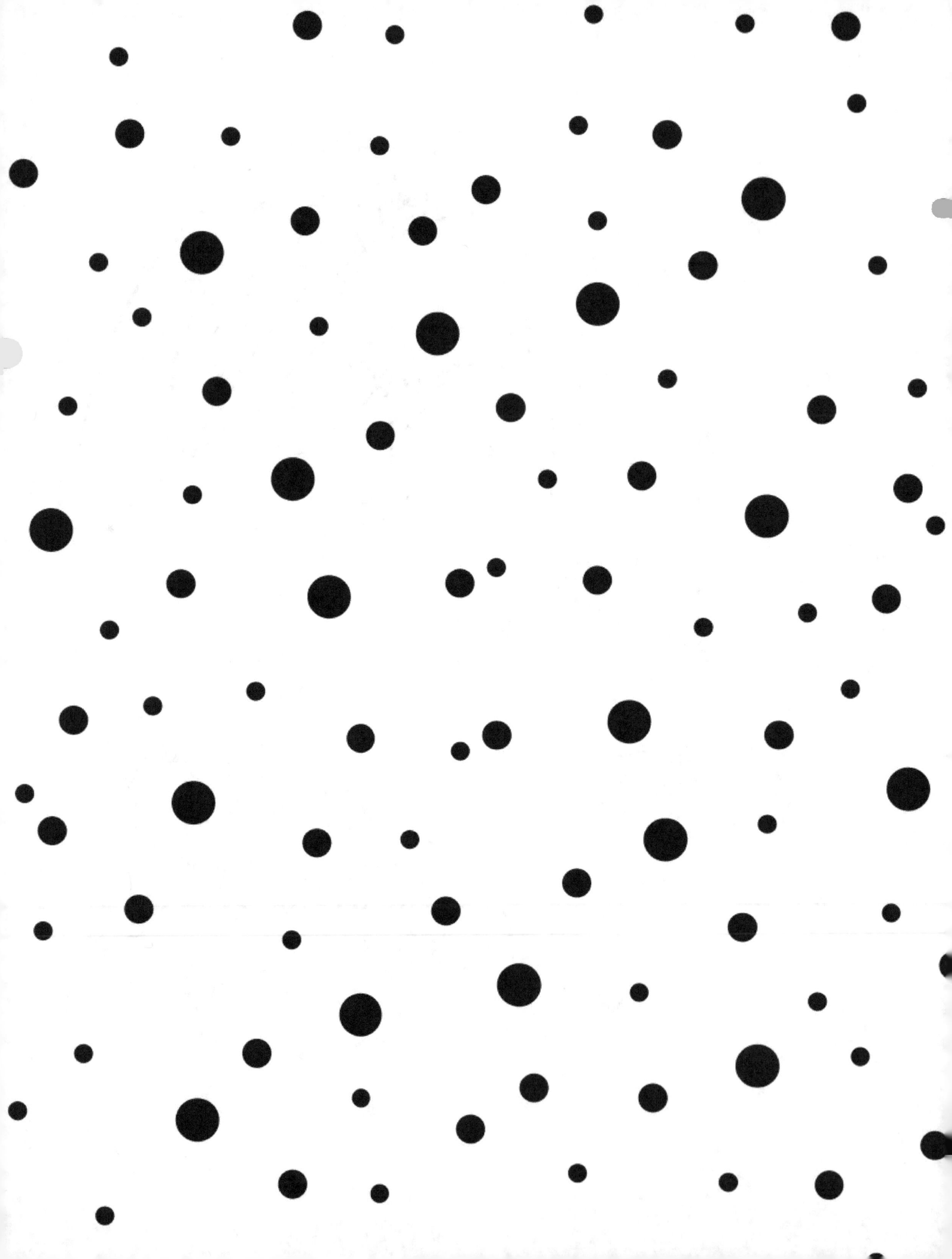

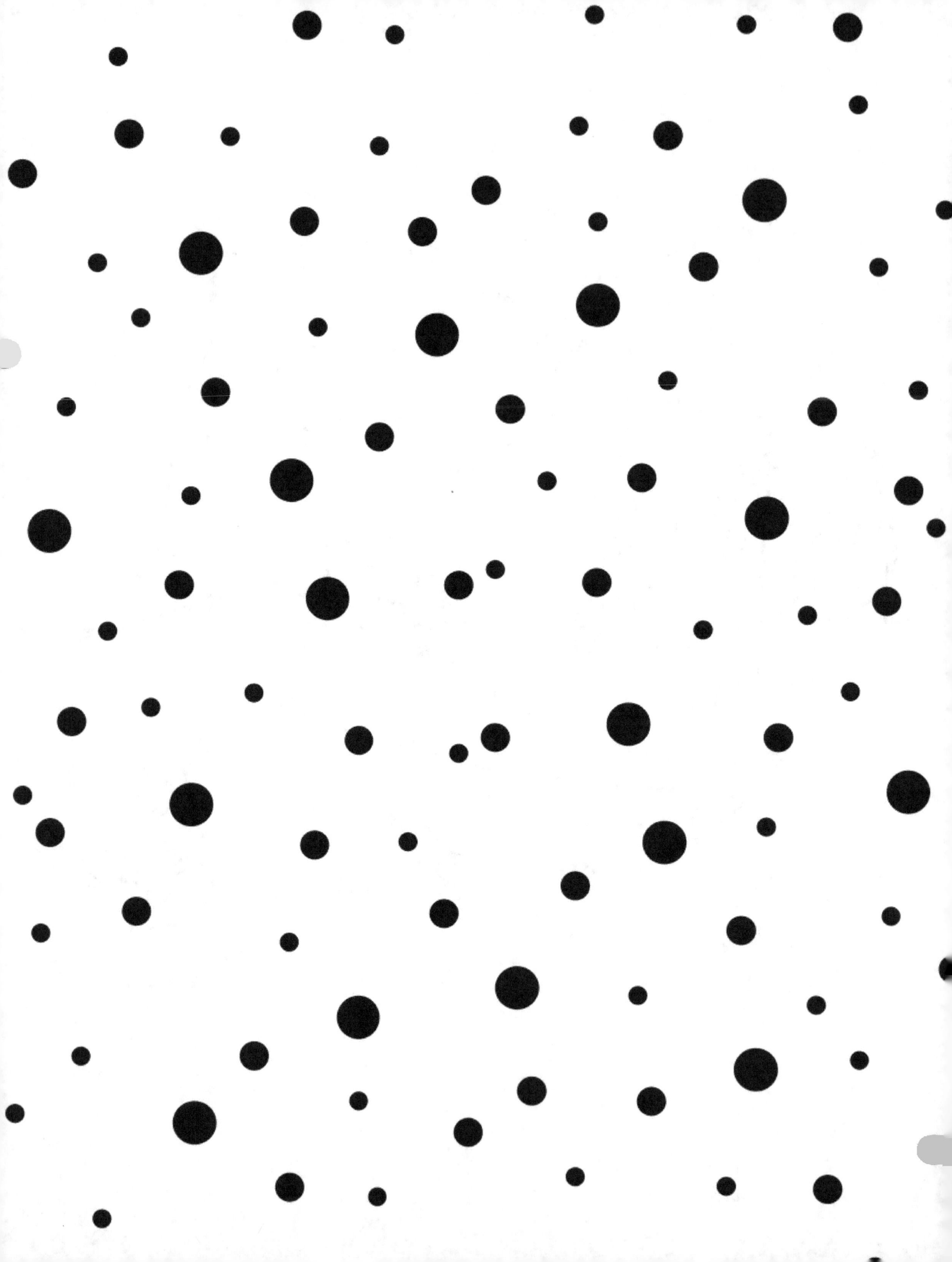

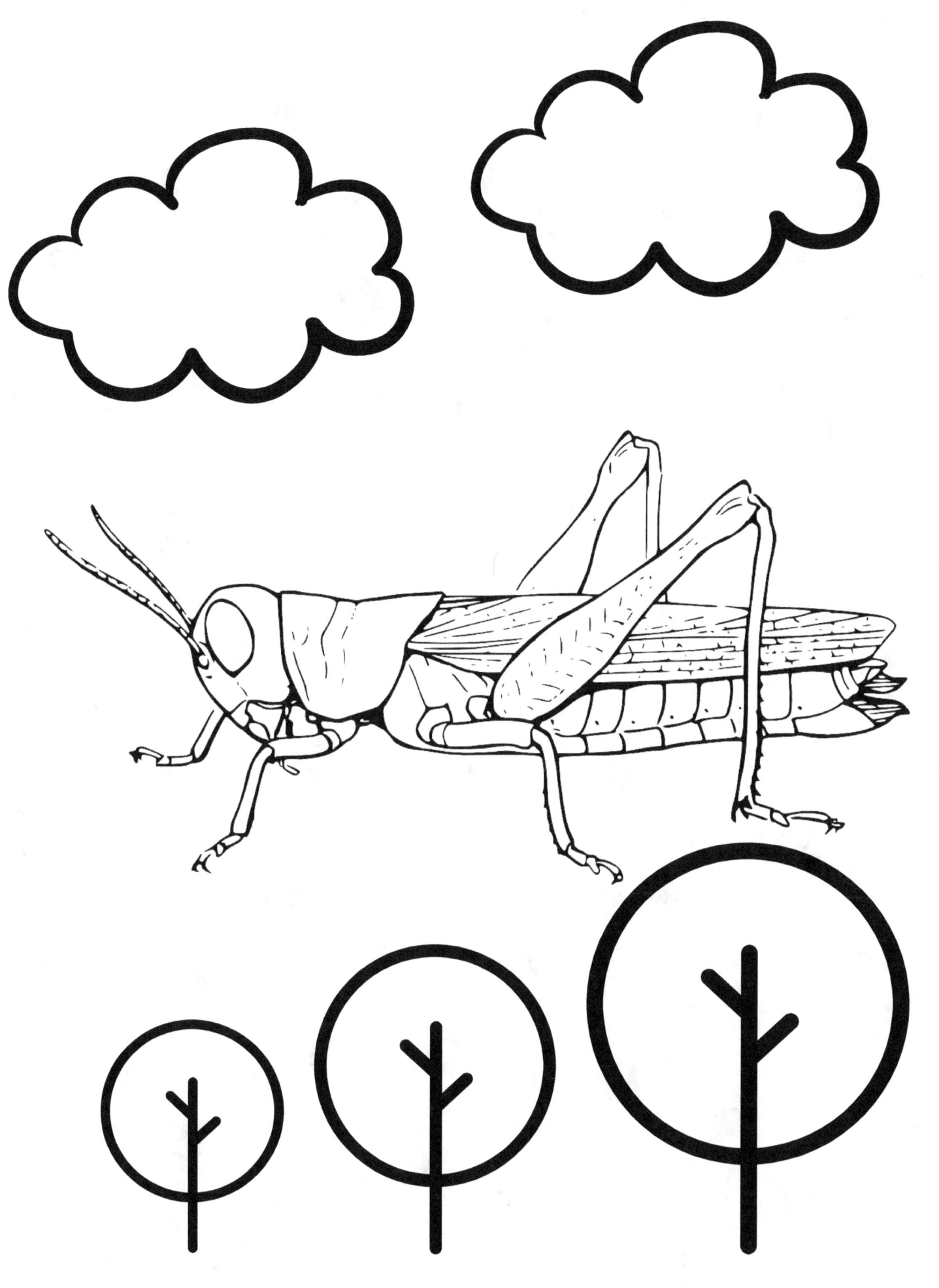

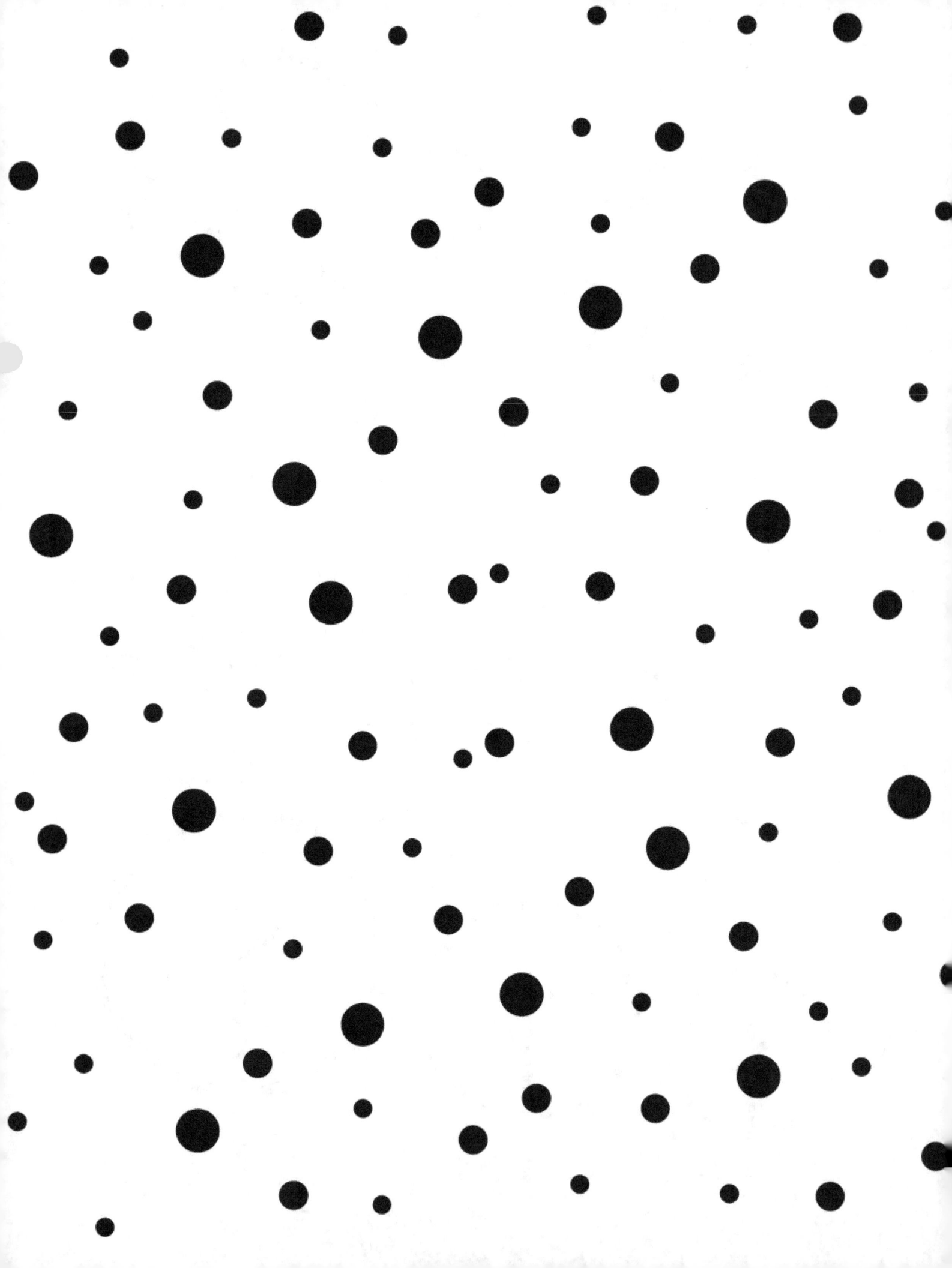

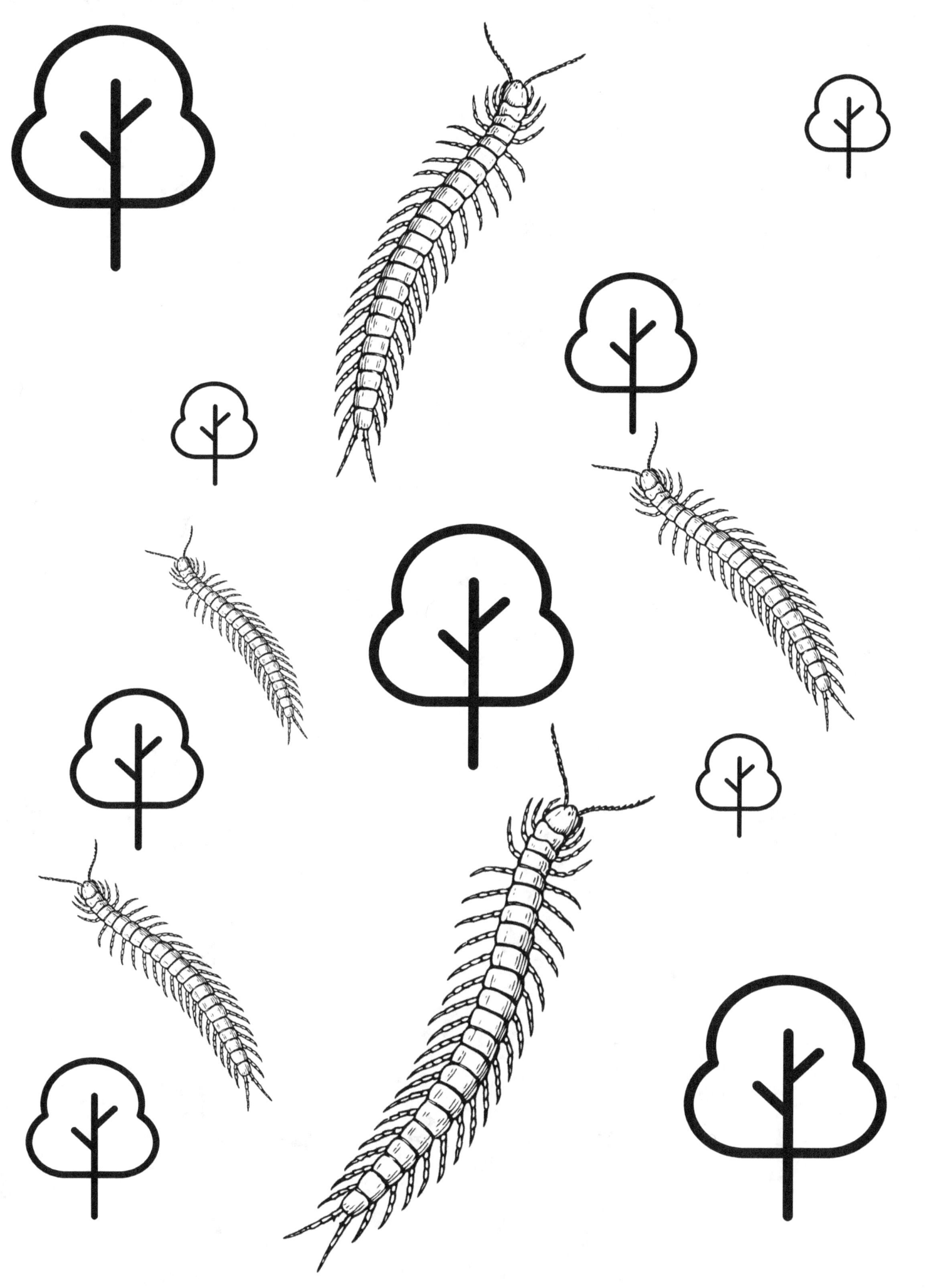

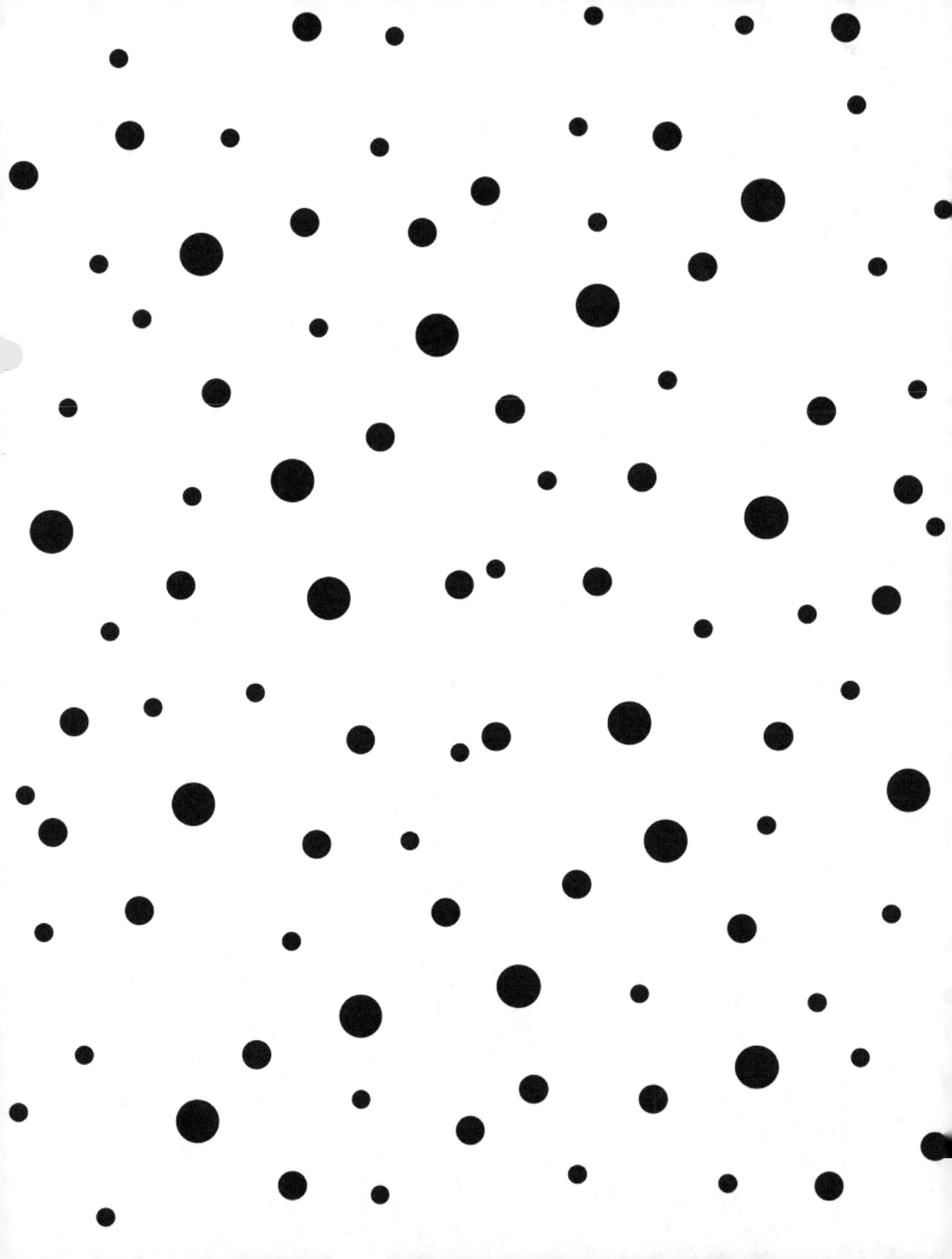

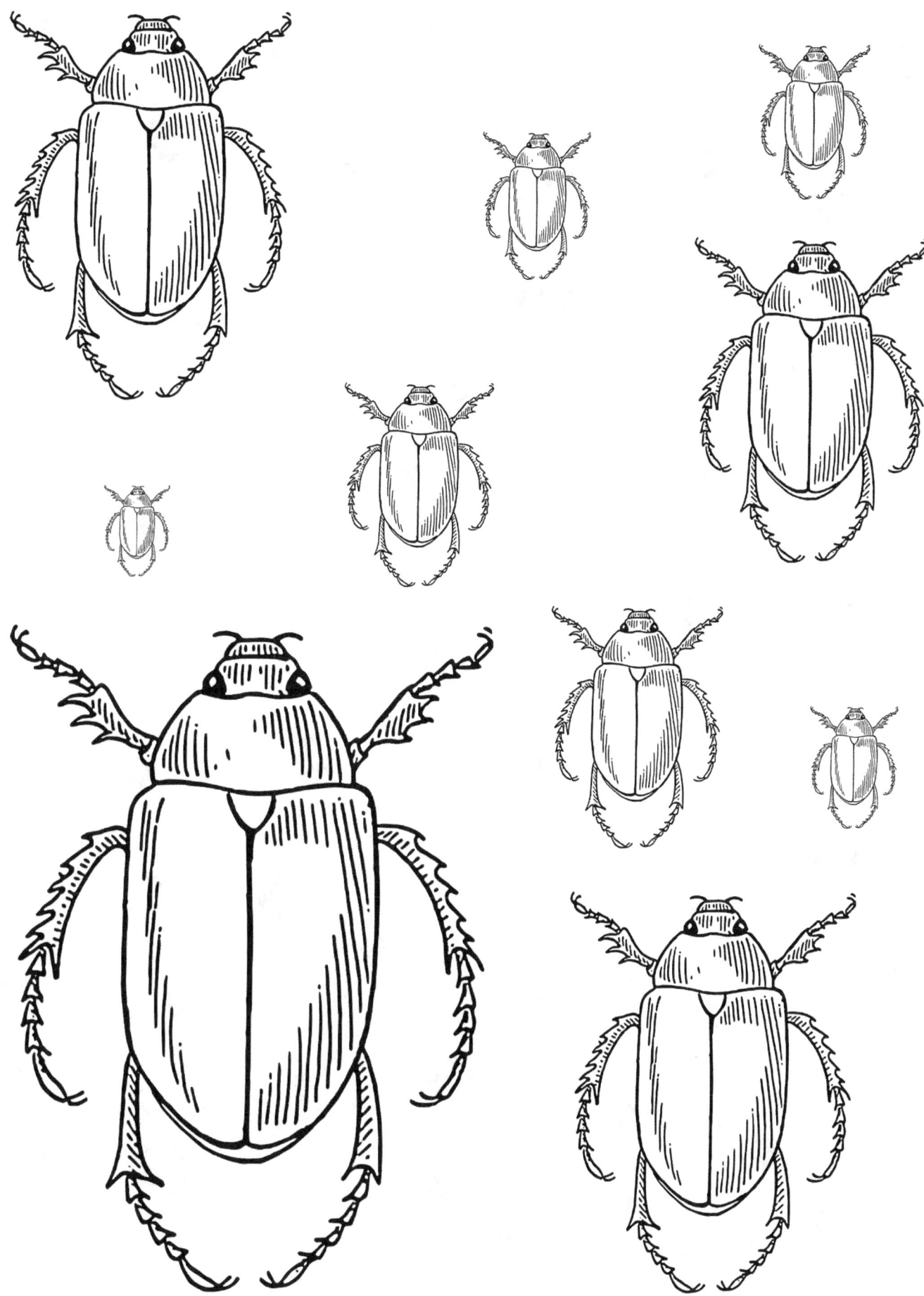

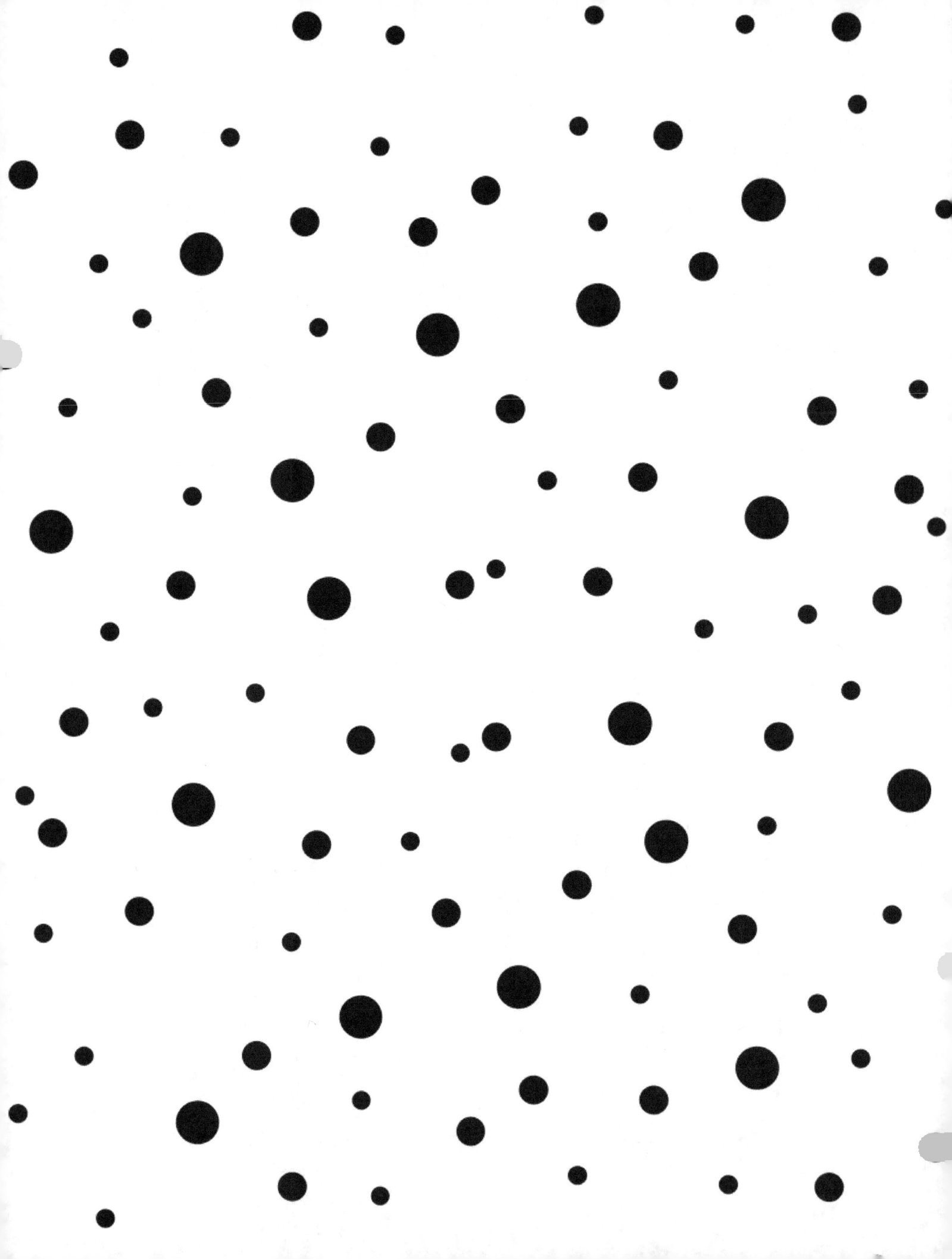

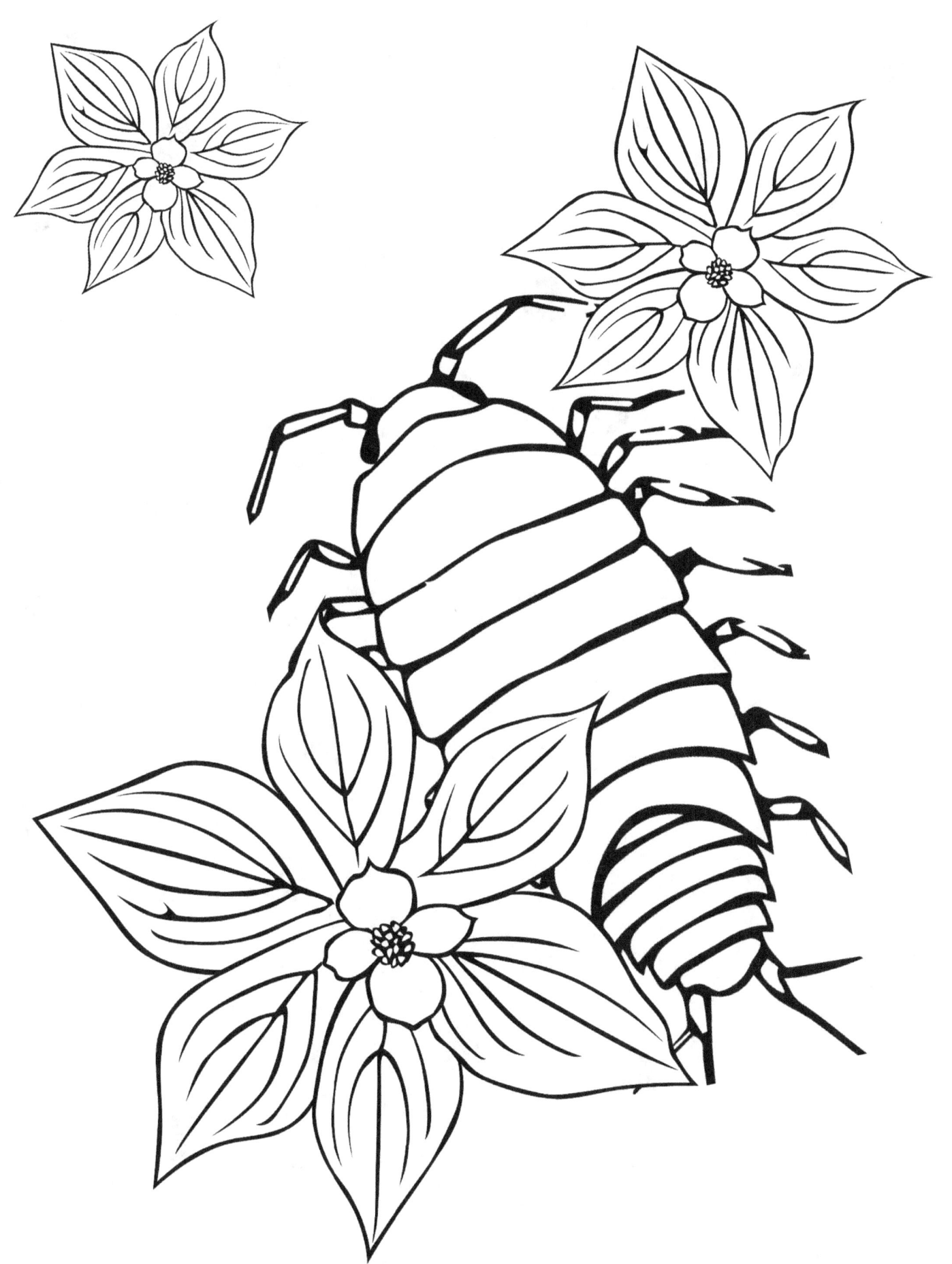

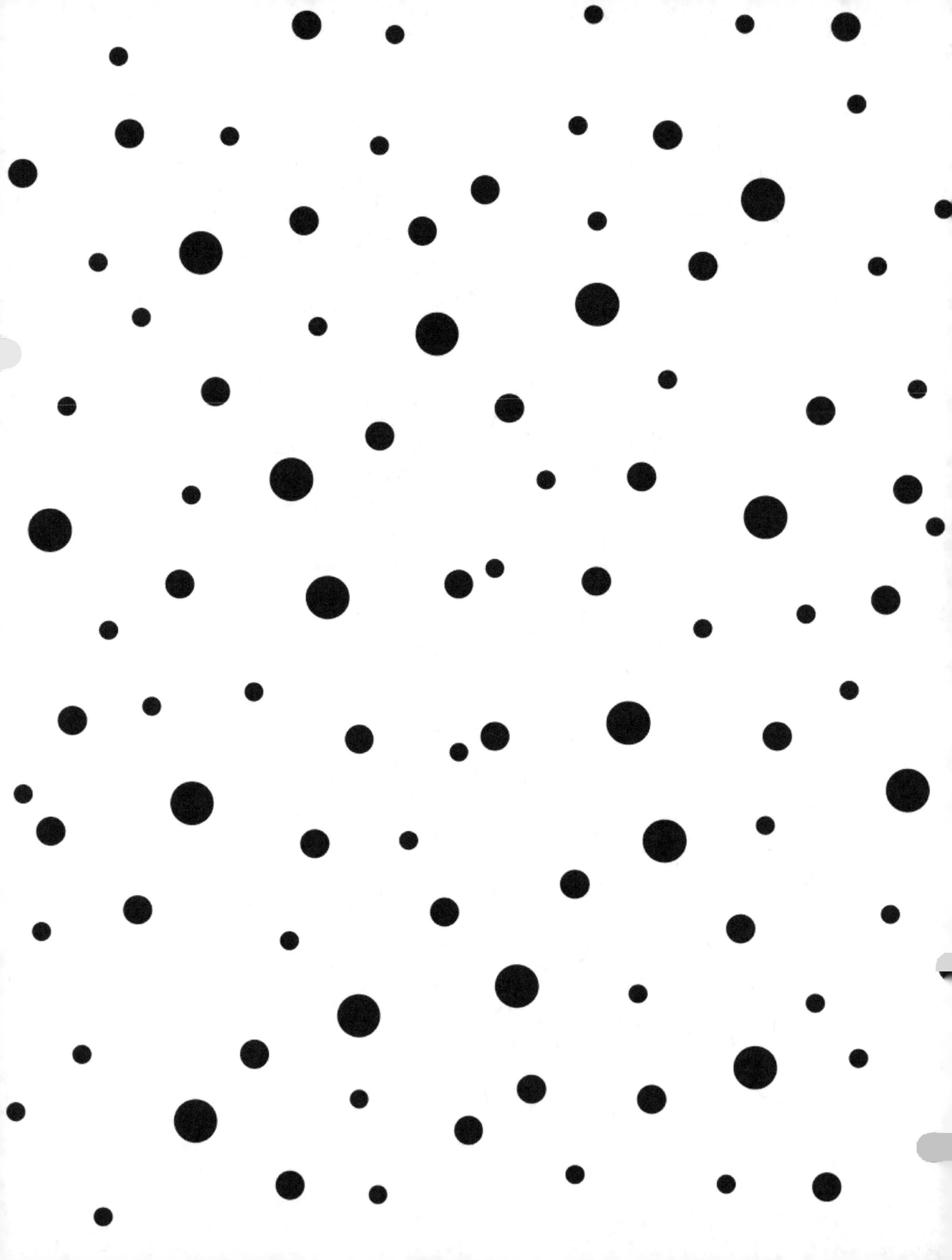

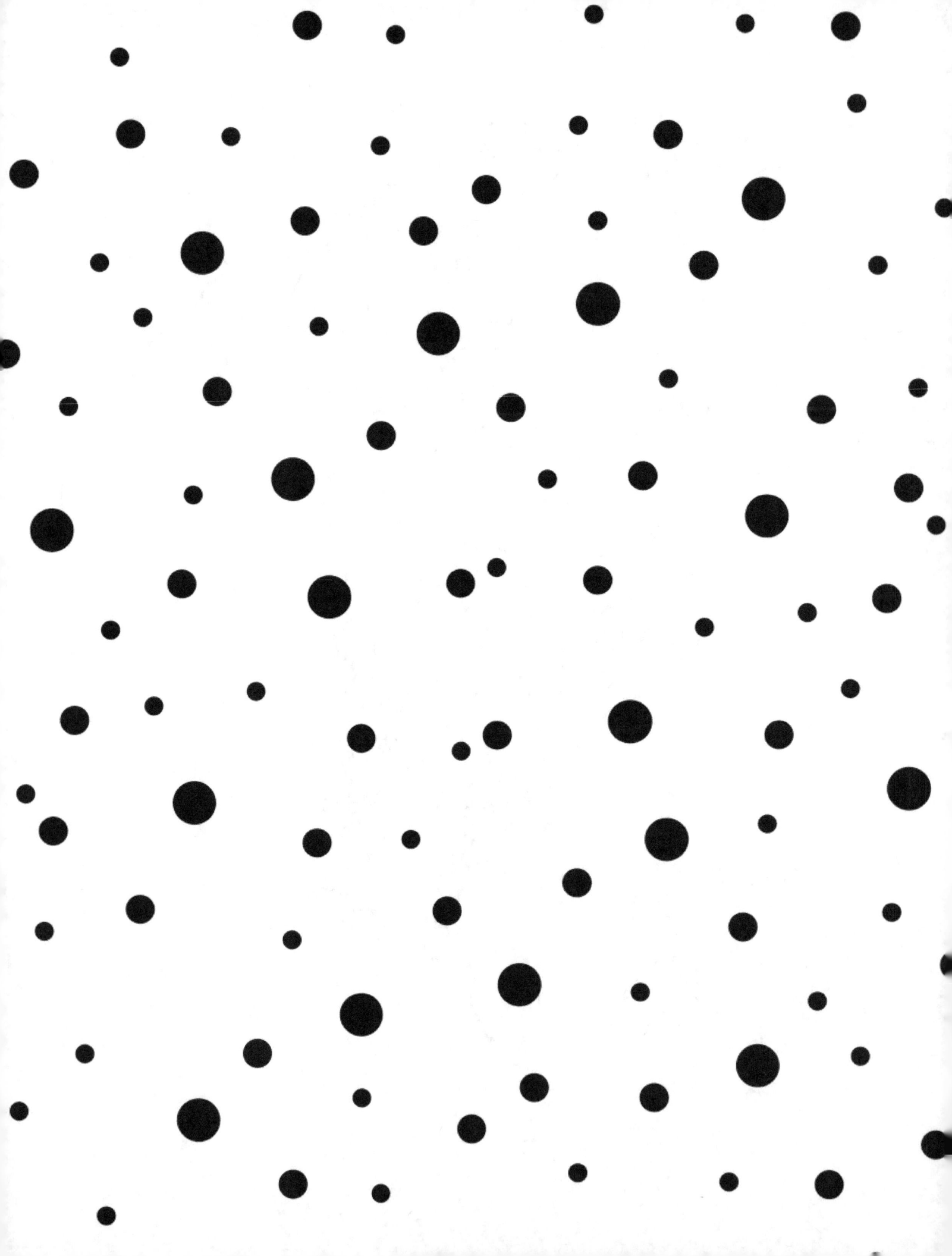

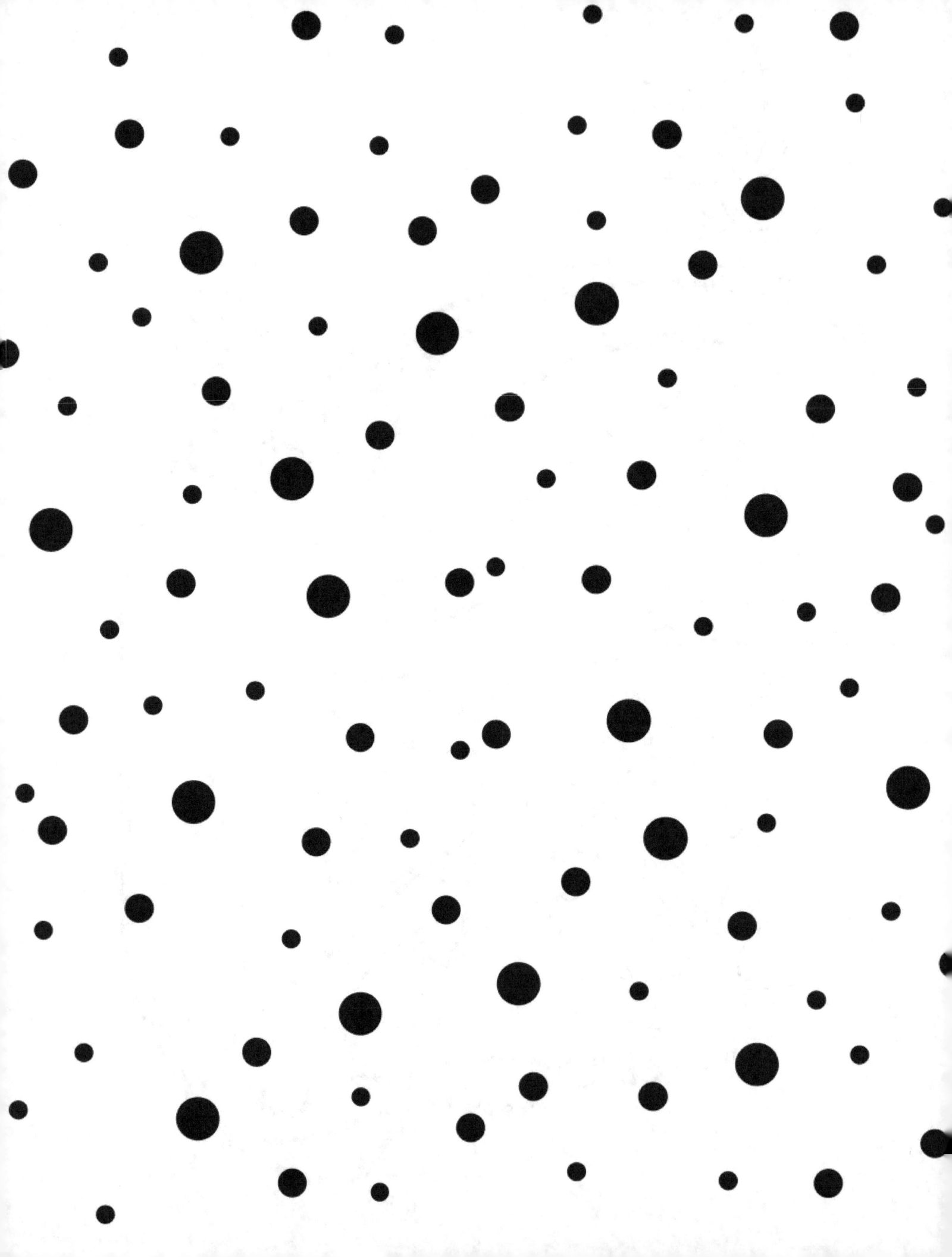

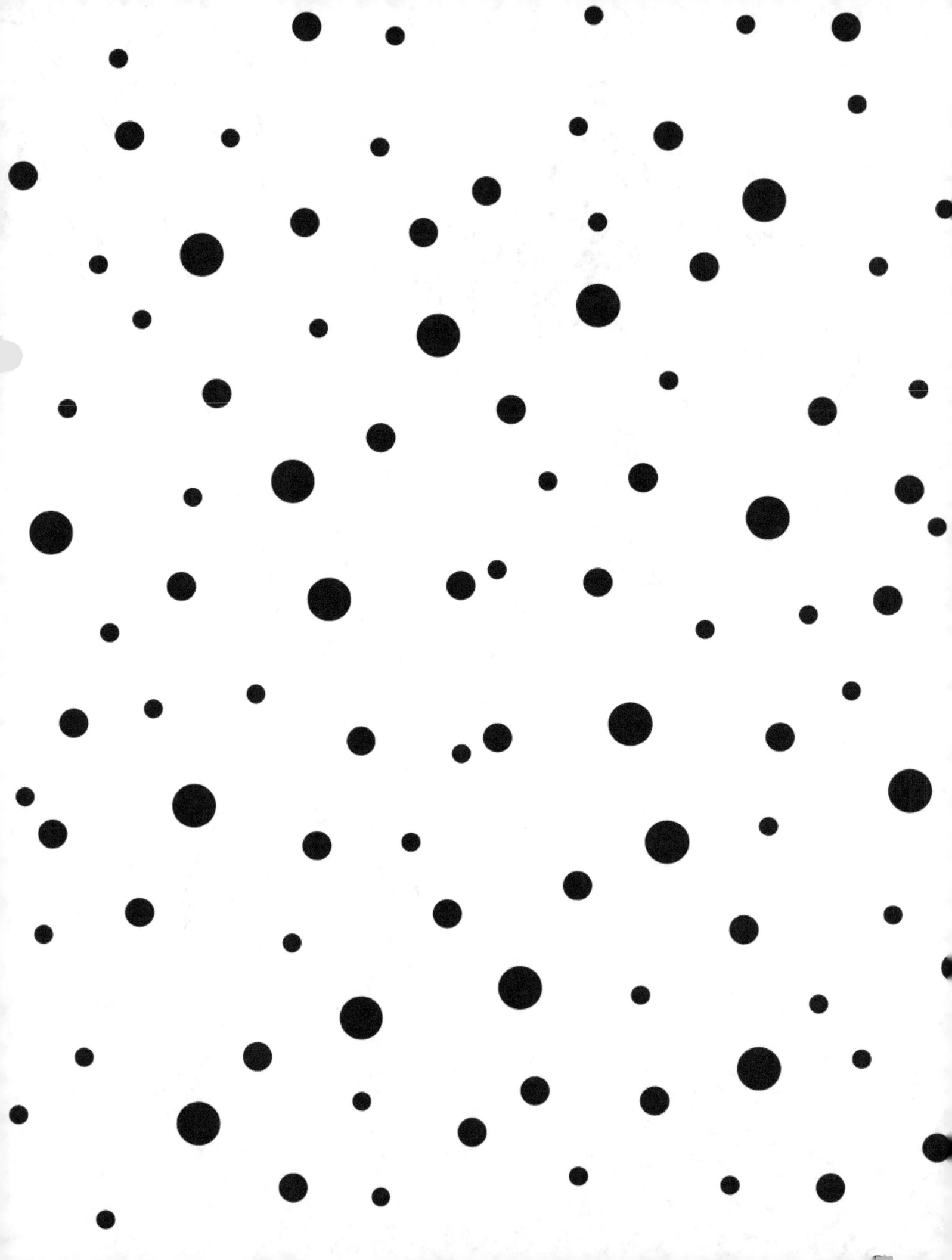

Thank You

OpenClipart-Vectors from Pixabay
Clker-Free-Vector-Images from Pixabay
yayang art from Pixabay

for many of the images that were used to create this book.
I appreciate your time, talent and energy.

www.ingramcontent.com/pod-product-compliance
Lightning Source LLC
Chambersburg PA
CBHW080951220526
45465CB00008BA/3247